COMPACT *Research*

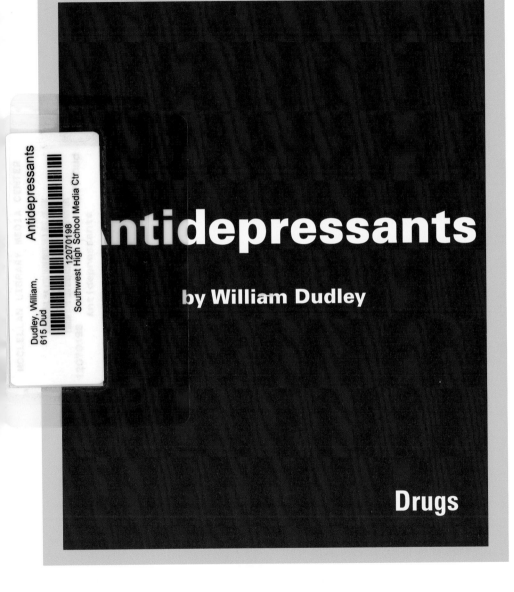

Antidepressants

by William Dudley

Drugs

ReferencePoint
Press™

San Diego, CA

© 2008 ReferencePoint Press, Inc.

For more information, contact:
ReferencePoint Press, Inc.
PO Box 27779
San Diego, CA 92198
www. ReferencePointPress.com

Picture credits:
Maury Aaseng: 37–40, 53–56, 70–73, 87–91
Istockphoto: 11
Landov:16

LIBRARY OF CONGRESS CATALOGING-IN-PUBLICATION DATA

Dudley, William, 1964–
   Antidepressants / by William Dudley.
      p. cm.
   Includes bibliographical references and index.
   ISBN-13: 978-1-60152-041-8 (hardback)
   ISBN-10: 1-60152-041-7 (hardback)
   1. Antidepressants—Juvenile literature. I. Title.
   RM332.D83 2008
   615'.78—dc22

                                                                    2007041602

Contents

Foreword

66 Where is the knowledge we have lost in information? 99

—"The Rock," T.S. Eliot.

As modern civilization continues to evolve, its ability to create, store, distribute, and access information expands exponentially. The explosion of information from all media continues to increase at a phenomenal rate. By 2020 some experts predict the worldwide information base will double every 73 days. While access to diverse sources of information and perspectives is paramount to any democratic society, information alone cannot help people gain knowledge and understanding. Information must be organized and presented clearly and succinctly in order to be understood. The challenge in the digital age becomes not the creation of information, but how best to sort, organize, enhance, and present information.

ReferencePoint Press developed the *Compact Research* series with this challenge of the information age in mind. More than any other subject area today, researching current events can yield vast, diverse, and unqualified information that can be intimidating and overwhelming for even the most advanced and motivated researcher. The *Compact Research* series offers a compact, relevant, intelligent, and conveniently organized collection of information covering a variety of current and controversial topics ranging from illegal immigration to marijuana.

The series focuses on three types of information: objective single-author narratives, opinion-based primary source quotations, and facts and statistics. The clearly written objective narratives provide context and reliable background information. Primary source quotes are carefully selected and cited, exposing the reader to differing points of view. And facts and statistics sections aid the reader in evaluating perspectives. Presenting these key types of information creates a richer, more balanced learning experience.

For better understanding and convenience, the series enhances information by organizing it into narrower topics and adding design features that make it easy for a reader to identify desired content. For example, in *Compact Research: Illegal Immigration*, a chapter covering the economic impact of illegal immigration has an objective narrative explaining the various ways the economy is impacted, a balanced section of numerous primary source quotes on the topic, followed by facts and full-color illustrations to encourage evaluation of contrasting perspectives.

The ancient Roman philosopher Lucius Annaeus Seneca wrote, "It is quality rather than quantity that matters." More than just a collection of content, the *Compact Research* series is simply committed to creating, finding, organizing, and presenting the most relevant and appropriate amount of information on a current topic in a user-friendly style that invites, intrigues, and fosters understanding.

Antidepressants at a Glance

Prevalence

Antidepressants are a big business, accounting for $20 billion in global sales in 2003. The majority of prescriptions for antidepressants are written by primary care physicians rather than psychiatrists or other specialists. One in 10 American women take antidepressants.

Types

Three basic categories of antidepressants are monoamine oxidase inhibitors (MAO: e.g., pargyline), tricyclic antidepressants (TCAs: e.g., amitriptyline, desmethylimipramine), and selective serotonin reuptake inhibitors (SSRIs: e.g., sertraline).

Pharmacological Properties

Most antidepressants have no perceptible effects on the central nervous system for people not suffering from depression. For people with depression, antidepressants can alleviate its symptoms, including feelings of sadness, lack of enjoyment, and fatigue. Some types of antidepressants (MAO inhibitors) may cause feelings similar to amphetamine after two to four weeks of use.

Laws

Antidepressants are not listed as restricted substances under the 1970 Controlled Substances Act. In the United States they are legally available only by prescription from a licensed medical doctor or psychiatrist. The Food and Drug Administration (FDA) must approve claims of their safety and effectiveness before they can be sold to the public.

Medical Uses

Antidepressants are primarily used to treat depression but have also been prescribed for other mood disorders including panic attacks, anxiety, eating disorders, and obsessive-compulsive disorders.

Side Effects

Side effects vary depending on the particular drug and may include sexual dysfunction, headaches, feelings of nervousness and anxiety, weight gain, and bone fragility. Some studies have linked antidepressant use with increased risk of suicide.

Chemical Dependency

Some studies have shown that up to one-third of antidepressant users who quit suffer from withdrawal symptoms such as dizziness, anxiety, and sensations in the body that feel like electric shocks. However, unlike opiates and sedatives, antidepressants are not considered addictive in that they do not create psychological cravings or require increased doses to maintain their effects.

Teens and Antidepressants

Teens and children have been prescribed antidepressants for depression, obsessive-compulsive disorder, chronic pain, anxiety, and other disorders; 2.1 million U.S. children under 12 and 8.1 million adolescents were prescribed antidepressants in 2002. The FDA in 2004 issued a health advisory against many popular antidepressants after a review of studies showed them linked to an increase in suicidal thoughts among young people.

Alternatives and Complements to Antidepressants

Natural and herbal alternatives to antidepressants are not regulated by the FDA. These include Saint-John's-wort, SAM-e, and ginkgo biloba. Also, many people undergo psychotherapy in addition to, or instead of, antidepressants to treat their depression.

Overview

The term *antidepressants* refers to a group of psychoactive medications that are used to treat depression and other mood disorders. Antidepressants do not have much effect on people who are not depressed. They do not give anyone any sort of "high" or "buzz." However, for people suffering from depression, antidepressants can reduce or eliminate the negative emotions and moods they are experiencing.

Antidepressants are widely used in the United States. Today they are frequently the first choice of psychiatrists, often in conjunction with some form of talk therapy, for treating people diagnosed with depression. Antidepressants are also frequently prescribed by family doctors for patients who are going through temporary stresses such as death of a loved one or divorce or are simply not feeling well. The popularity of antidepressants as a therapeutic tool is a relatively recent phenomenon that resulted from the development in the 1980s and 1990s of new types of antidepressants that had fewer side effects. An estimated 20 million Americans, including 1 million children and adolescents, take antidepressants.

How Do Antidepressants Work?

Research into the causes of depression and other mood disorders and how antidepressants alleviate the symptoms of depression is still ongoing. But most scientists agree that antidepressants work by interacting with neurotransmitters—chemicals in the brain that bind with neurons (brain cells), enabling them to communicate with each other. Some studies have suggested that depression is related to having too few neurotransmitters available to the brain, reducing its electrochemical activity. The key neurotransmitters affected by antidepressants include serotonin, dopamine, and norepinephrine. Serotonin is believed to carry messages in the brain related to feelings of well-being and calmness. Dopamine influences a person's motivation and perception of reality; people in depressed moods often have low dopamine levels. Norepinephrine plays a role in the body's response to stress.

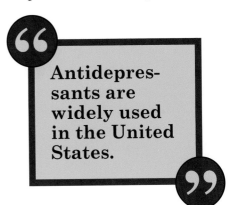

Antidepressants are widely used in the United States.

Antidepressants change the levels of these neurotransmitters in the brain. Some do so by blocking reuptake—the natural process by which neurotransmitters are absorbed and eventually recycled in the body. Others interfere with the rate of neurotransmitter breakdown or prevent neurotransmitters from chemically binding with certain nerve cells. All these processes can leave more neurotransmitter cells available for nerve cell messaging within the brain.

Types of Antidepressants

Antidepressants are generally classified into categories that describe how they change the chemistry of the brain. Two types of antidepressants—MAO inhibitors and tricyclics or TCAs—were developed in the 1950s. While effective for some patients, they did not attain widespread use due to adverse side affects. In the 1980s a new type of antidepressants called selective serotonin reuptake inhibitors (SSRIs) was approved for use in the

United States. Heavily marketed by pharmaceutical companies under such brand names as Prozac, Paxil, and Zoloft, these pills quickly became popular and supplanted talk therapy as the leading therapeutic tool for depression. More recently drug companies have come up with new generations of drugs that work with different combinations of neurotransmitters.

> "Antidepressants change the levels of . . . neurotransmitters in the brain."

MAO Inhibitors

Monoamine oxidase inhibitors (MAO inhibitors or MAOIs) were accidentally discovered in the 1950s by scientists looking for a drug to treat tuberculosis. They noticed that some patients given a tuberculosis drug called iproniazid experienced greatly elevated mental moods. In a 1957 article from *Better Homes and Gardens*, an anonymous 24-year-old man describes how iproniazid changed his mood. "I felt like 60, now I feel like 15. I can't get depressed any more even when I try."[1]

MAO inhibitors work by preventing the enzyme monoamine oxidase from breaking down neurotransmitters in the brain. This enables greater numbers of neurotransmitter cells to remain in the brain, stimulating neuron activity.

However, MAO inhibitors can also cause serious side effects. The original MAO inhibitor, iproniazid, was removed from the market after it was shown to cause severe liver damage to patients. Other MAO inhibitors were developed that avoided this particular problem, but in the early 1960s several patients died because of mysterious side effects including headaches, high blood pressure, and brain hemorrhages. It was later found that monoamine oxidase also breaks down tyramine, a chemical that helps regulate blood pressure. Inhibiting its effects can create fatal increases in blood pressure.

People on MAO inhibitors must therefore avoid a long list of foods and medicines because of how they interact with the drug. Restricted foods include beer, wine, certain cheeses, chocolate, fava beans, ripe figs, yogurt, and beverages with caffeine. Restricted drugs include common cold and cough medicines, nasal decongestants, and other antidepressants. People taking MAO inhibitors often carry medical alert bracelets so that in cases of medical emergency, the doctors will have that information.

Antidepressants are frequently the first choice of psychiatrists, often in conjunction with some form of talk therapy, for treating people diagnosed with depression. An estimated 20 million Americans, including 1 million children and adolescents, take antidepressants.

Because of these side effects and related food and medicine restrictions, MAO inhibitors were never widely used in the United States (they were more popular in Great Britain). MAO inhibitors drugs that remain on the market today include phenelzine (Nardil) and tranylcypromine (Parnate).

Tricyclic Antidepressants

Tricyclics, or TCAs, get their name from the three-ring carbon chemical structure they share. Like MAO inhibitors, this group of antidepressants was a somewhat accidental discovery of the 1950s. Drug manufacturers were investigating a class of drugs called antihistamines, which were used (and are still used today) to treat nasal congestion. While looking for a type of antihistamine that did not cause drowsiness, they found one that was effective for the mental illness of schizophrenia. This in turn led to the discovery of a related chemical (imipramine) that relieved symptoms of depression. Imipramine is still sold under the brand name Tofranil; other tricyclics being prescribed today include amitriptyline (Elavil), desipramine (Norpramin), and nortriptyline (Aventyl or Pamelor).

> Two types of antidepressants—MAO inhibitors and tricyclics or TCAs—were developed in the 1950s.

Tricyclics work by inhibiting the reuptake of two brain neurotransmitters—serotonin and norepinephrine. Depression symptoms are thus eliminated by the increase of functioning neurotransmitters in the brain. However, tricyclics also affect other chemicals, such as histamine and acetylcholine, that in turn affect other parts of the body. Imipramine, like other TCAs, "is a dirty drug—a drug that affects many systems at once,"[2] writes psychiatrist and author Peter D. Kramer. Side effects are therefore common. These range from irritating (dry mouth and eyes, upset stomach, lightheadedness) to more serious (weight gain, sexual dysfunction, blurred vision). TCAs may also aggravate heart problems or other medical conditions. In addition, an overdose of tricyclics can be fatal. Writing in 2006, author Suzanne LeVert states that "overdoses of TCAs are the leading cause of overdose deaths in the United States, and have been for at least a decade."[3] Taking them requires careful monitoring. Because of these and other complications, tricyclics—the leading class of antidepressant prescribed in the United States from the 1960s to the 1980s—never attained widespread popularity.

SSRIs

First introduced in the United States in the late 1980s, SSRIs (selective serotonin reuptake inhibitors) are the most widely prescribed and heavily studied class of antidepressants. Many have become household names, including fluoxetine (Prozac), sertraline (Zoloft), and paroxetine (Paxil). Other notable SSRIs include citalopram (Celexa) and fluvoxamine (Luvox).

As their name suggests, SSRIs work on one neurotransmitter—serotonin—and have little effect on others. By preventing the reuptake (chemical absorption) of serotonin, SSRIs increase the level of this particular neurotransmitter in the brain. In addition, some scientists believe that boosting the level of serotonin may stimulate the brain to increase its amount of serotonin receptors—specialized molecules on neuron cells that enable them to bind with serotonin.

SSRIs are not necessarily more effective in treating depression than the older-generation tricyclics and MAO inhibitors, but they are generally recognized as having far fewer side effects. SSRI side effects include nausea or agitation when first taking the drug and sexual dysfunction for some patients. Some SSRI antidepressants have been linked to teen suicide.

> "First introduced in the United States in the late 1980s, SSRIs (selective serotonin reuptake inhibitors) are the most widely prescribed and heavily studied class of antidepressants."

When Prozac and other SSRIs were introduced, they quickly became a cultural phenomenon—far more so than the earlier MAO inhibitors and tricyclics. News of their effectiveness spread both through word of mouth and through mass media coverage, including cover stories in *Time* and *Newsweek* and a best-selling book by psychiatrist Peter D. Kramer. Antidepressants were prescribed for millions of people, not all of whom were undergoing treatment for depression. Writing in 1996, health writer Nancy Wartik noted that by then six out of 10 antidepressant prescriptions were being dispensed by family practitioners rather than mental health specialists. Antidepressants, she wrote, were being "prescribed for a far greater range of ailments and for less serious disorders; whereas tricyclics were once reserved only for those with severe depression, these days it's not uncommon for physicians to prescribe Prozac for a case of the blues."[4]

New-Generation Antidepressants

Spurred by the enormous economic success of Prozac and other SSRIs, scientists at pharmaceutical companies are continually developing new types of antidepressants that change brain chemistry but with minimal side effects. Some of these drugs fall into the categories previously described. For example, new forms of MAO inhibitors, called reversible MAO inhibitors, are safer than traditional MAO inhibitors because their side effects can be more quickly reversed. But many of these newer drugs do not fall readily into any of the three categories. One popular antidepressant of this sort is bupropion (Wellbutrin), which affects the reuptake of both serotonin and dopamine, and which is marketed as having fewer sexual side effects than SSRIs. Another antidepressant, venlafaxine (Effexor) inhibits the reuptake of both serotonin and norepineprhine.

Other newer antidepressants do not affect the reuptake of neurotransmitters, but instead affect select nerve cell receptors—the molecules on neurons that bind with and take messages from neurotransmitters. This action is believed to affect brain chemistry and may help relieve depression symptoms. Mirtazapine (Remeron) affects the nerve cell receptors that take messages from norepinephrine. Yet other antidepressants have a dual action in that they affect both reuptake and nerve cell receptors. An example is nefazodone (Serzone) which inhibits the reuptake of serotonin and blocks a certain type of serotonin receptor.

Uses of Antidepressants

Depression

The treatment of depression is a main use of antidepressants. Depression is a mental illness characterized by long and debilitating feelings of sadness, despair, fatigue, and diminished interest in eating and other everyday activities. These emotions are sometimes severe enough to impair people's ability to function or do routine tasks. In extreme cases, they may lead to suicide. In addition to emotional symptoms, depressed people also often suffer from physical symptoms such as exhaustion and low energy, changing sleep patterns, headaches, and problems with eating too much or too little. The belief that depression is simply a character flaw or personal weakness remains popular among many Americans, but psychologists and other mental health professionals have declared depression a mental illness of the brain that can and should be treated.

For much of the twentieth century depression was primarily treated with counseling and talk therapy overseen by psychologists. Such therapies were often time-consuming, expensive, and slow to achieve concrete results. Since the 1980s antidepressants have become the number one therapeutic tool for treating this illness. For 70 percent of patients, antidepressants can effectively relieve the emotional and physical symptoms of depression, although they do not necessarily solve the underlying causes of the disease. In some cases, if an antidepressant does not work initially, a doctor can change the dosage or the type of antidepressant until one is found that works. Typically, patients use antidepressants for four to six months, often in conjunction with therapy. In some cases, however, patients and doctors may decide to use antidepressants for longer periods of time.

> " For 70 percent of patients, antidepressants can effectively relieve the emotional and physical symptoms of depression. "

The rising popularity of antidepressants such as Prozac in the 1990s coincided with a large increase in the number of patients being diagnosed with depression even if they did not show all the clinical symptoms. The number of Americans being diagnosed for depression tripled between 1980 and 2004. Some of this increase may be due to a decrease in social stigma attached to depression, but many believe these changes were an effect of the burgeoning popularity of antidepressants and the people's demand for a quick fix. Medical historian Edward Shorter wrote in 1998 that because of antidepressants such as Prozac, depression itself had been transformed from "an unusual condition that led to hospitalization" to "a common malady for which help [antidepressants] was readily available."[5]

In addition to depression, other mental and physical problems are also treated with antidepressants.

Bipolar Disorder

Also known as manic depression, bipolar disorder is characterized by extreme mood swings. People with this disease oscillate between periods of depressive lows and manic highs. These periods of mania may take

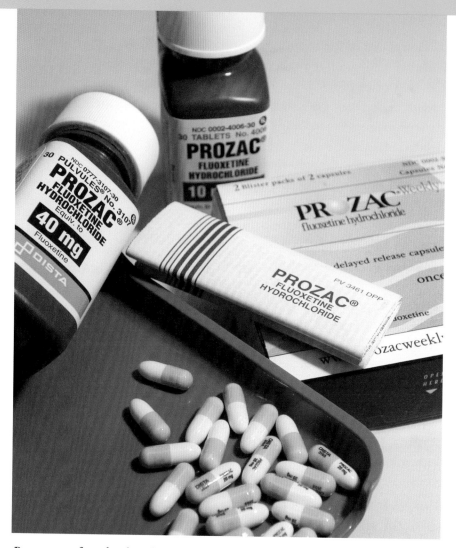

Prozac was first developed as a chemical in 1974. It took years of testing before the FDA approved it for sale as an antidepressant in 1987. According to a 2000 investigation by the Indianapolis Star *newspaper, Eli Lilly paid more than $50 million to quietly settle more than 30 lawsuits alleging that Prozac created suicidal or violent behavior.*

the form of unusual cheerfulness and enthusiasm but may also result in extreme irritability or temper tantrums. Some patients—especially young ones—experience "mixed episodes" of intense mania and depression simultaneously.

Bipolar disorder is most often treated with lithium or other drugs called mood stabilizers (anticonvulsant drugs have been found to be especially effective). However, in some cases antidepressants are also prescribed to help patients deal with periods of depression. But because antidepressants can often trigger or cause manic episodes, they should be used with care and possibly in conjunction with mood stabilizers.

Anxiety Disorders

Anxiety disorders are mental illnesses in which anxiety and fear become overwhelming and interfere with a person's life. They occur in approximately 17 percent of the U.S. population. Many people who suffer from depression also suffer from anxiety. Types of these illnesses include social anxiety disorder (disabling fears of being in social situations), panic attacks (sudden acute attacks of fear, including physical symptoms of pounding heart, sweating, and difficulty breathing), and obsessive-compulsive disorder (being plagued by intrusive recurring thoughts and compulsive behaviors).

Doctors sometimes prescribe SSRI antidepressants such as fluoxetine (Prozac) or sertraline (Lustral) for obsessive-compulsive disorder, and citalopram (Cipramil) for panic attacks. Paxil was marketed for some time as a treatment for social anxiety disorder. Trazodone (Molipaxin), a tricyclic antidepressant, is sometimes prescribed for anxiety, while moclobemide (Manerix), an MAO inhibitor, is sometimes tried for patients with social phobias.

Chronic Pain

Some antidepressants are effective pain relievers, especially for people with chronic pain. They work regardless of whether the patient has depression. Amitriptyline—a tricyclic antidepressant—is the one most heavily studied and commonly prescribed. It and other tricyclics help people with the burning or painful sensations that accompany nerve damage from diabetes, shingles, or strokes. It is thought that they work by increasing the amount of neurotransmitters in the spinal column that suppress pain messages. Newer antidepressants that work on both serotonin and norepinephrine, including venlafaxine (Effexor) and duloxetine (Cymbalta) are also sometimes prescribed for patients with pain.

Chemical Dependency Treatment

Some types of antidepressants have been prescribed for nicotine addiction and similar problems. For example, bupropion (Wellbutrin) has been prescribed under the brand name Zyban for people attempting to quit smoking.

Regulation of Antidepressants

Antidepressants are medications available only by prescription. The primary government agency in the United States that has authority over antidepressants is the Food and Drug Administration (FDA), an agency of the U.S. Department of Health and Human Services. Founded in 1906, the agency's original purpose (which has not changed) was to ensure that foods were unadulterated with toxins or germs and that the labeling of ingredients in foods and drugs was accurate. In 1962 Congress passed legislation giving the FDA the power to withhold drugs from the market until the sellers provide scientific studies demonstrating their drugs' safety and effectiveness. These studies are generally performed or funded by the drug manufacturers themselves.

> "The primary government agency in the United States that has authority over antidepressants is the Food and Drug Administration.

Based on the information they receive, the FDA not only has the power to withhold or remove drugs from the market but also to classify drugs as to whether they can be sold "over the counter" or solely with a doctor's prescription. The FDA can also regulate how drugs are advertised and mandate the placement of warnings and information if necessary. In March 2004, based in part on public testimony and consumer complaints, the FDA mandated that some antidepressants carry a warning label stating that they may increase the risk of suicide in teens and that patients taking them must have close supervision.

One interesting quirk about America's regulatory system is that once a medication is approved for the market for a certain disease and for certain patients, doctors may legally prescribe the medication for purposes and types of patients that have not been reviewed or approved by the FDA. This practice is called "off-label" prescribing. For example, of the major antidepressants, only fluoxetine (Prozac) has been studied among adolescents and approved for adolescent use. But some psychiatrists and doctors have issued their teen patients "off-label" prescriptions for other antidepressants that have received FDA approval only for adult use. In other cases a drug that may have FDA approval only for treating depression might be prescribed for obsessive-compulsive disorder or other conditions.

Concerns About Antidepressants

Despite their legal status, FDA regulation, and widespread popularity, antidepressants are still the subject of ongoing questions and controversies. These include questions about their long-range safety and effectiveness, their use by young people, and whether they are being overprescribed.

How Safe and Effective Are Antidepressants?

Despite numerous testimonials on the effectiveness of antidepressants, some mental health experts have expressed doubts about their safety and effectiveness. Some critics argue that the FDA is not doing enough to ensure that antidepressants are truly effective and safe before it approves them. They argue that drug companies submitting research to the FDA present only the studies that support their drugs, while not releasing any studies that indicate otherwise. Some studies have found that antidepressants are no more effective in alleviating depression symptoms than a placebo pill. Antidepressants may work simply because people believe they can, based on what they have learned from pharmaceutical company advertisements and the media.

> " Despite their legal status, FDA regulation, and widespread popularity, antidepressants are still the subject of ongoing questions and controversies. "

Concerns have also been raised about the safety of popular antidepressants. Side effects have been reported even among SSRIs that are touted as being relatively clean of such problems. These include both physical and psychological symptoms, which can include *akathisia*, feelings of intense anxiety and jumpiness that may trigger rage or hostility. Some patients attempting to quit taking antidepressants have reported withdrawal symptoms ranging from dizziness, tremors, and diarrhea to suicidal thoughts.

Should Antidepressants Be Prescribed to Children and Adolescents?

Worries about the safety of antidepressants have been especially acute regarding children and adolescents. About 5 percent of antidepressant prescriptions are given to patients under 18. The brains and bodies of children often absorb and/or react to drugs in ways different from adults, and few antidepressants have been specifically approved for use for young people.

> Some patients attempting to quit taking antidepressants have reported withdrawal symptoms ranging from dizziness, tremors, and diarrhea to suicidal thoughts.

In December 2003 a British medical body advised that teens not be prescribed antidepressants because they were linked with increased risks of suicide. The FDA has not taken antidepressants off the market for teens, but in 2004 it did mandate a "black box" warning label that stated that antidepressants may trigger severe mood swings, aggression, and hostility for people 18 and under and that teens taking antidepressants should be closely monitored for increased suicide risk.

Are Antidepressants Overprescribed?

Concerns about the safety and effectiveness of antidepressants for teens and adults have led some to speculate whether physicians are overprescribing antidepressants. In many cases family practitioners and other nonpsychiatrists have responded to patients' complaints about "sadness" by prescribing antidepressants, often with little consultation, monitoring, or therapy. These doctors, with less training in mental health issues than psychiatrists, may be confusing clinical depression with normal feelings of sadness triggered by life events.

For some, ethical and philosophical issues remain over whether a person should be encouraged to change his or her personality through chemicals—even legally prescribed chemicals. Widespread prescription of antidepressants leaves the impression that any grieving or sadness is a medical problem to be treated with a drug, rather than a life problem to be resolved and worked through by the individual.

Are Alternatives to Antidepressants Effective?

People worried about taking antidepressants, or patients who have been taking them and want to stop, have tried various alternatives to anti-depressants. These include herbal remedies such as Saint-John's-wort, ginkgo biloba, and nutritional supplements such as omega-3 fatty acids or SAM-e. Proponents tout these as safe and natural alternatives that can relieve depression with fewer side effects than drug antidepressants. But supplements and herbal remedies are not under as strict scrutiny by the FDA as antidepressants and do not have to pass the same test hurdles to be sold.

In addition to these medicines, people have pursued traditional talk therapy, electrical stimulation of the brain, and various alternative treatments such as hypnosis and acupuncture.

The Continuing Importance of Antidepressants

Antidepressants, while not as celebrated as they were in the early 1990s, remain an important tool for mental health practitioners and doctors. Most Americans will either try antidepressants at some time or know someone who has. The more people know about these drugs, the better they can decide whether to use them, and how they can use them effectively.

How Safe and Effective Are Antidepressants?

> **❝150 years from now future generations may look back on the antidepressant era as a frightening experiment.❞**

—David A. Karp, *Is It Me or My Meds? Living with Antidepressants.*

> **❝Antidepressant drugs are not happy pills, and they are not a panacea. . . . They are, however, one depression treatment option . . . and there is evidence that they do help.❞**

—John Hauser, "Depression Medications: Antidepressants."

Tracy Thompson, a staff writer for the *Washington Post,* called her depression "The Beast." She would lie awake for hours, not wanting to "get up and face another day of deadening duties and crushing sadness." Her depression left her isolated from her friends and family. "I felt trapped in an invisible, airless chamber, my frantic gestures ignored or misunderstood. To others, I merely seemed remote and angry."

Then, after years of seeking treatment through therapy and medication, in 1990 she tried a new antidepressant drug called Prozac. Within a few days she felt her anxiety recede and her warring feelings inside her head cease. "It was as if the evil beast that had been holding my head under water, trying to drown me, had suddenly let go." Three years later, she was still taking the antidepressant and was enjoying her "chemically altered" state. "I don't feel sedated, jittery, or drugged. I simply feel normal—as if I had been driving a car all these years with the parking brake on, and now it is off. I feel as if the real me has returned."[6]

Differing Experiences with Antidepressants

Susan L., a graphic designer in Manhattan, switched from an older antidepressant to Prozac when it was introduced in the late 1980s. For five years Prozac worked as promised to alleviate sad feelings and other depression symptoms. Then the drug lost its effectiveness. She has since tried various other combinations of antidepressants. "I tried Zoloft, I tried Effexor. I was on Wellbutrin and Effexor for a while, now I'm on Wellbutrin and Celexa," she told a newspaper reporter in 2002. "It's still not that great. I couldn't say that I'm happy."[7]

Susan, a teacher and mother in Oakland, California, sought treatment from a psychiatrist for depression and anxiety while coping with menopause, a dying mother, and a failing marriage. She began taking the antidepressant Paxil when her therapist suggested it would ease her feelings and make her therapy more productive. At first she liked it. The pills "worked for me" by making her "comfortably numb." But then her physical health began to deteriorate. She became incontinent, developed an immune disorder, and was constantly dizzy. After three years of being diagnosed with various syndromes and cancers, she looked up the documented side effects of Paxil and recognized all of her symptoms. She stopped taking Paxil, only to have a large withdrawal reaction. Then, with another doctor's help, she gradually tapered off her dosage of the antidepressant. Three weeks after her last dose, her physical symptoms vanished. "Of course, I still battle with some anxiety and mood swings," she told a newspaper reporter. "But I am me again—yes, more anxious, but physically healthy and mentally clear. And I am a lot happier."[8]

Tim "Woody" Witczak, age 37, had started a new job in 2003; its demands were causing anxiety and insomnia. His doctor prescribed the antidepressant Zoloft. He soon began suffering from agitation, nightmares, and strange physical sensations, such as feeling his head was detached from his body. Five weeks later, he hung himself. "He had no history of mental illness," asserted his wife Kim, who believes that Zoloft induced the suicide and that its manufacturer, Pfizer, knew of this side effect risk but kept it from the public.[9]

Debating the Safety and Effectiveness of Antidepressants

All of these stories illustrate a fundamental fact about antidepressants—they affect different individuals in different ways. They also provide anecdotal evidence—positive and negative—on the general safety and effectiveness of antidepressants. The safety and effectiveness of these drugs has been a matter of debate ever since Prozac and other SSRI-type antidepressants exploded on the American scene in the late 1980s.

The Food and Drug Administration

The government agency in the United States responsible for ensuring that prescription medicines are safe and effective is the Food and Drug Administration (FDA). The FDA requires that before a drug is sold in the United States, clinical tests must be done that demonstrate its safety and efficacy. Prozac was first developed as a chemical in 1974. It took years of testing before the FDA approved it for sale as an antidepressant in 1987.

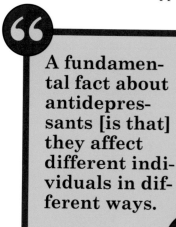

A fundamental fact about antidepressants [is that] they affect different individuals in different ways.

The clinical tests required by the FDA for antidepressant and other drug approval are medical experiments that follow established procedures (called protocols). Tests generally use human volunteers who are placed into different groups. One group is given the experimental antidepressant (or whatever medicine is being studied). Another group—the control group—is sometimes given a standard treatment, such as an antidepressant already in common use, or other treatment such as therapy that is generally recognized as a viable treatment for depression. After a period of time the two groups are compared to see if the group receiving the experimental medication attained better results for their condition than the control group.

In other clinical studies, the research subjects in the control group receive a placebo—a pill or medicine with inactive ingredients. For example, one group of research subjects would take a blue pill that contained an experimental antidepressant, while the other group would take a blue sugar pill. The experimental antidepressant would have to be statistically more successful than the placebo in order to justify further studies and ultimate FDA approval.

How Effective Are Antidepressants?

All antidepressants sold in the United States have gone through a battery of clinical tests and other studies to gain FDA approval. The body of data collected on antidepressants and reviewed by the FDA has created a general consensus within the health care industry on their general effectiveness. All types of antidepressants—tricyclics, SSRIs, and others—are roughly equal in their success rates in treating depression (they differ mainly in their side effects). They all have a 60 percent to 80 percent rate of improving a person's mood and alleviating depression symptoms. All types of antidepressants generally do not have an instant effect on a person's mood (they do not produce a "high" or "buzz") but often instead take several weeks to work. All types of antidepressants should be taken with caution and under a doctor's care. Doctors cannot predict with certainty whether a particular antidepressant will work on a particular person. Many times a person may try two or more antidepressants or antidepressant combinations before finding one that works.

> " All antidepressants sold in the United States have gone through a battery of clinical tests and other studies to gain FDA approval. "

Side Effects

In addition to determining whether a certain drug is effective, clinical tests also help discern the drug's side effects. Research test subjects are carefully observed to see what side effects they are experiencing. When the FDA approves a drug for prescription it often requires that information about the possible side effects be provided to the doctor and consumer (this provides much of the fine print one sees on medical labels). In many cases these effects are mild and disappear after taking the drug for a week or two. But some documented instances of antidepressant side effects are life-threatening.

MAO inhibitors can cause agitation, dry mouth, weight changes, and sexual side effects. They may also cause a person's blood pressure to rise rapidly by lowering the body's supply of tyramine, a body chemical that regulates blood pressure. This in turn may trigger strokes, heart attacks, and other serious complications. A person taking these antidepressants must avoid many types of foods, including certain cheeses, red wine, coffee, chocolate, eggplant, and aged meat. Because of these side effects and complications, doctors and psychiatrists in the United States have generally avoided prescribing MAO inhibitors unless other antidepressants have been tried and have failed to work.

The side effects of tricyclic antidepressants, the mainstay of depression medicine from the 1960s and 1980s, are numerous and often unpleasant. They include feelings of restlessness and anxiety, difficulty concentrating, dry mouth and eyes, constipation, difficulty urinating, nausea, and dizziness. They may worsen certain medical conditions such as heart disease or an enlarged prostate. Because of these side effects, many patients who tried them often quit taking the drugs before they could be effective in helping their depression. Another safety problem with tricyclics is the fact that an overdose could be fatal. The physician or psychiatrist must be extremely careful in monitoring doses and managing side effects when prescribing tricyclics.

SSRIs and Newer Antidepressants

SSRIs, introduced in the late 1980s and later, were first marketed as being superior to older antidepressants because they had fewer side effects and because they did not have the fatal overdose potential of tricyclics. But they are not free of side effects. These can include stomach upset, weight changes, nervousness, headaches, and sexual problems, including loss of sexual desire and inability to have an orgasm. SSRI drugs may also cause bone loss among elderly patients.

A few patients may have a more serious reaction to SSRI antidepressants. These drugs work by increasing the levels of the neurotransmitter serotonin in the brain. But too much serotonin in a person's system may cause "serotonin syndrome" whose symptoms include hallucinations, confusion, and the inability to control muscular movements. If left untreated—or if a patient is misdiagnosed and is given increased doses of the SSRI antidepressant—serotonin syndrome can cause death via stroke or heart attack.

Another problem associated with SSRI antidepressants is akathisia—extreme drug-induced agitation. Some have described akathisia's effects as an injection of itching powder inside one's skin. It can also create paranoia and rage and has been blamed for extreme violent acts, including violent suicides.

The success of SSRI antidepressants such as Prozac, Zoloft, and other SSRIs has caused drug companies to look for and develop drugs that work in similar ways but with fewer side effects. Some of the new drugs were SSRIs, but other newer antidepressants work in a variety of different ways and target different combinations of neurotransmitters, neurotransmitter subtypes, and cellular activities of the brain. These new-generation antidepressants have on average worked about as well as SSRIs and have roughly the same general constellation of possible side effects.

Questioning Antidepressants

The general consensus on antidepressants approved by the FDA—that they can be effective for the majority of patients, although their side effects can cause safety concerns—has been challenged by some patients and mental health practitioners. Concerns range from whether they are as effective as advertised to whether the FDA process has been compromised and has allowed dangerous drugs to be prescribed to Americans.

The Placebo Effect

One of the complications in determining the actual effectiveness of antidepressants (and other medications) is the placebo effect. Placebos are the fake medicines given to participants in a clinical drug study. In many cases patients respond favorably to the placebo. Most researchers agree that this is the result of a patient *believing* that he or she has taken potent and helpful medicine. The belief itself "fools" the brain and body and creates a positive result, such as a decrease in depression symptoms. The placebo response could also be the result of the increased attention depressed people receive when they are participants in a clinical study.

> One of the complications in determining the actual effectiveness of antidepressants (and other medications) is the placebo effect.

Some people believe that the placebo effect works not only with placebos but with antidepressants as well. In other words, the positive mental results people describe from taking antidepressants are actually a result of the placebo effect—a self-fulfilling prophecy of positive thinking. Antidepressants seem to be successful solely because people strongly believe that the drug they are taking has to work; and bolstered by this positive confidence, they find themselves improving their emotional well-being. Even experiencing unpleasant physical side effects can be interpreted by some as proof that the antidepressants must be working, reasoning on a conscious or subconscious level that if such a drug is powerful enough to cause dizziness or sexual dysfunction, it must be working on their depression or anxiety as well. But if a placebo works just as well as an antidepressant for many, critics argue, then any side effect or risk of antidepressants should count as a reason to discourage their use.

Are Antidepressants Addictive?

Another question surrounding antidepressants is whether they cause addiction or chemical dependency. According to the American Academy of Family Physicians, antidepressants fundamentally differ from such drugs as tranquilizers and barbiturates precisely because "they aren't addictive."[10] Antidepressants are not addictive in the sense that people feel a craving for them if they stop taking them. However, a proportion of antidepressant users (estimates vary from 5 to 60 percent) experience withdrawal symptoms if they stop taking antidepressants. These symptoms may include headaches, dizziness, fatigue, nausea, and vomiting.

Another question surrounding antidepressants is whether they cause addiction or chemical dependency.

Some patients also report feeling sensations like electric shocks from inside their heads. To avoid this, people who quit antidepressants often gradually reduce their dose rather than stop cold turkey.

Is the FDA Doing Its Job?

Many critics of antidepressants are also critics of the FDA. Several problems have been noted. One is that drug companies are not required to submit the results of all their studies to the FDA or even release them to the public. If the drug company has done two tests, one of which indicated a drug was effective and another one that did not, it could submit just the positive one for review. Another problem is that too little research and monitoring is done once a drug is approved. "While the FDA demands strict data on efficacy and safety from clinical trials before approving a new drug," notes *Washington Post* reporter Shankar Vedantum, "less attention is paid after the drug reaches the market."[11] Thus, a problem that may not show up in clinical trials but later shows up in 1 of every 2,000 people could result in thousands of cases of serious side effects if the drug is approved and becomes as popular as many antidepressants have.

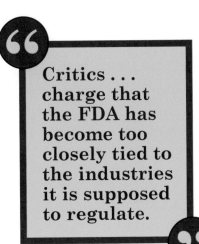

Critics . . . charge that the FDA has become too closely tied to the industries it is supposed to regulate.

Critics also charge that the FDA has become too closely tied to the industries it is supposed to regulate. This could be the result of the fact that much of the agency's budget comes from user fees from the pharmaceutical companies, who then negotiate how these fees are spent. Furthermore, many of the scientific advisory panels that the FDA uses to provide recommendations are peopled by scientists who have consulting arrangements or other financial ties to these companies. "It has become clear that the FDA is too closely tied to the industries it represents and is unable to protect the public from unsafe drugs," argues Congressman Maurice Hinchey, citing antidepressants as one example of "how the FDA failed to quickly respond to numerous warning signals and put millions of Americans at risk."[12] The Pharmaceutical Research and Manufacturers of America has responded to criticism and calls for reform by noting that fewer than 3 percent of approved prescription drugs have been withdrawn from the American market for safety reasons over the last 20 years. "Though there is always room for improvement, it would be a mistake to accept that the FDA drug safety system is seriously flawed."[13]

How Safe and Effective Are Antidepressants?

66 My concern is if you list every possible side effect [of antidepressants], people will say, 'I'll have all of these? Then, I shouldn't take it.' 99

—Julio Licinio, quoted in Lidia Wasowicz, "Antidepressants Often Not Effective, Healthy, Safe, or Good for Children," UPI, June 17, 2006.

Licinio is a psychiatry professor at the University of California at Los Angeles.

66 There are no documented long-term adverse effects from antidepressants. 99

—Ava T. Albrecht and Charles Herrick, *100 Questions & Answers About Depression*. Sudbury, MA: Jones and Bartlett, 2006.

Albrecht is a practicing psychiatrist in New York City and a member of the faculty at the New York University School of Medicine. Herrick is medical director for Intensive Psychiatric Services at Danbury Hospital in Danbury, Connecticut, and a faculty member at New York Medical College.

Bracketed quotes indicate conflicting positions.

* Editor's Note: While the definition of a primary source can be narrowly or broadly defined, for the purposes of Compact Research, a primary source consists of: 1) results of original research presented by an organization or researcher; 2) eyewitness accounts of events, personal experience, or work experience; 3) first-person editorials offering pundits' opinions; 4) government officials presenting political plans and/or policies; 5) representatives of organizations presenting testimony or policy.

❝Industry controls the data, and industry with the aid of FDA have miscoded the data so all the articles in all the journals that purport to represent clinical trial data [on antidepressants] is misleading.❞

—David Healy, quoted in Shankar Vedantum, "Antidepressants a Suicide Risk for Young Adults," *Washington Post*, December 14, 2006.

Healy is a psychiatrist, author, and a prominent critic of antidepressants and the pharmaceutical industry.

❝SSRIs may cause a general soothing of anger and irritability over day-to-day events. This is an effect separate from depression.❞

—Edward Drummond, *The Complete Guide to Psychiatric Drugs*. Hoboken, NJ: John Wiley & Sons, 2006.

Drummond is associate medical director at the Seacoast Mental Health Center in Portsmouth, New Hampshire.

❝The side effects [of antidepressants] are immediate, but the beneficial effects take some time.❞

—Thomas Schwartz, quoted in Marianne Szegedy-Maszak, ". . . but Still Sad; Antidepressants Aren't the Magic That Millions Hoped," *Los Angeles Times*, March 27, 2006.

Schwartz is a professor of psychiatry at the State University of New York (SUNY) Medical University in Syracuse.

❝I am me again—yes, more anxious, but physically healthy and mentally clear. And I am a lot happier.❞

—Susan, quoted in Marianne Szegedy-Maszak, ". . . but Still Sad; Antidepressants Aren't the Magic That Millions Hoped," *Los Angeles Times,* March 27, 2006.

Susan, a teacher and mother, took the antidepressant Paxil for several years to combat anxiety and depression but stopped taking it after she developed unpleasant physical symptoms.

❝There's definitely a reluctance by the FDA to come out and say, 'A drug we've approved is really dangerous.'❞

—Richard Kapit, quoted in David Stipp, "Trouble in Prozac," *Fortune,* November 28, 2005.

Kapit is a former FDA official who oversaw the safety review of Prozac prior to its 1987 FDA approval; he now works as a medical writer and consultant.

❝The suicides under SSRIs are violent. . . . It's jumping, knives, pain. They're in pain. They're jumping out of their skins.❞

—Vera Sharav, quoted in Richard C. Morais, "Prozac Nation? Is the Party Over?" *Forbes,* August 20, 2004.

Sharav is president of the Alliance for Human Research Protection, a group that contends that the patients and clinical trial volunteers have not been fully informed about the risks of antidepressants.

❝We think everybody at the time they are started on antidepressants ought to be observed closely because it's a dangerous time.❞

—Robert Temple, FDA press briefing, December 13, 2006.

Temple is director of the FDA's Office of Medical Policy.

....

❝I think so-called antidepressants are just drugs that do other things, like sedating or stimulating people.❞

—Joanna Moncrief, interviewed by Reuters Health, July 15, 2005.

Moncrief is a senior lecturer at the University College London's Department of Mental Health Services.

....

❝Depression patients treated with antidepressants have shown improved performance on various neurocognitive measures, compared with untreated depression patients.❞

—Lakshmi Ravindran and Sidney H. Kennedy, "Are Antidepressants as Effective as Claimed? Yes, but . . ." *Canadian Journal of Psychiatry,* February 2007.

Ravindran is a medical doctor who has done graduate studies of psychiatry at the University of Toronto in Ontario, Canada. Kennedy is a professor of psychiatry at the University of Toronto.

....

❝Although serotonin poisoning can be caused by an antidepressant overdose, it more often results from a combination of an SSRI or MAOI with another serotonin-raising substance.❞

—Jane E. Brody, "A Mix of Medicines That Can Be Lethal," *New York Times,* February 27, 2007.

Brody is a medical writer and columnist for the *New York Times.*

....

"We have no way to know beforehand which patient will respond to which antidepressant. And we have no way to know which, if any, side effects a particular person will develop."

—Francis Mark Mondimore, *Adolescent Depression*. Baltimore: Johns Hopkins University Press, 2002.

Mondimore is a practicing psychiatrist and member of the clinical faculty at the Johns Hopkins University School of Medicine.

"Medications for depressive disorders are not habit forming."

—National Institute of Mental Health, *Men and Depression*. Bethesda, MD: NIMH, 2005.

The National Institute of Mental Health (NIMH) is a federal government agency that oversees government policy and research on mental health issues.

"Anyone who has been on an antidepressant for more than about a month can experience withdrawal symptoms and dependence if the drug is stopped abruptly."

—Joseph Glenmullen, *The Antidepressant Solution*. New York: Free Press, 2005.

Glenmullen is a practicing psychiatrist, a clinical instructor at Harvard Medical School, and the author of several books including *Prozac Backlash*.

66 The . . . adverse event effects of SSRI drugs make their use hardly justified in the majority of cases because SSRIs are well known to have limited efficacy over placebo and . . . non-pharmacologic treatments. 99

—Donald Marks, testimony before an FDA advisory panel on antidepressants, February 3, 2004.

Marks is a physician who previously worked as director of clinical research for two multinational pharmaceutical companies.

66 In practice, in the treatment of major depression, I almost always combine medicine with psychotherapy. I imagine that . . . the antidepressants allow for the learning that psychotherapy guides. 99

—Peter D. Kramer, *Against Depression.* New York: Viking, 2005.

Kramer is a clinical professor of psychiatry at Brown University and the author of several books including *Listening to Prozac.*

66 As a physician, I was taught that all medications have risks and side effects so that I must consider the risk-to-benefit ratio of any medication I prescribe. Here the benefits [of antidepressants] far outweigh the risks. 99

—Paula Clayton, statement before an FDA advisory panel on antidepressants, December 13, 2006.

Clayton is a psychiatrist and medical director of the American Foundation for Suicide Prevention.

Facts and Illustrations

How Safe and Effective Are Antidepressants?

- **Neurotransmitters are chemicals involved in the sending of messages between neurons**, the basic cells of the brain and nervous system. They include serotonin, dopamine, and norepinephrine. Most antidepressants are believed to work by affecting the patient's levels of neurotransmitters in the brain.

- **Thirty percent** of depressed patients do not respond to the first antidepressant they try.

- About **65 percent** of people who do not respond to or are not helped by one type of antidepressant do respond to another.

- It is recommended that people helped by antidepressants **should continue taking them at least six months** after their depression recedes.

- According to a 2000 investigation by the *Indianapolis Star* newspaper, Eli Lilly paid **more than $50 million** to quietly settle more than 30 lawsuits alleging that Prozac created suicidal or violent behavior.

- Beginning in 2005 the FDA has issued several public advisories warning **pregnant women not to take Paxil and other SSRI antidepressants** because of possible harm to the baby.

- According to a 2005 FDA public health advisory, **instances of suicidal behavior among adults treated by antidepressants** may have increased.

How Brain Cells Send Messages to Each Other

Brain cells communicate via neurotransmitters with the following steps: (A) Neurotransmitters are released into the synapse (gap) between the sending and receiving brain cells. (B) A neurotransmitter binds with a receptor on the receiving nerve cell, altering its activity. (C) The neurotransmitter is released back into the synapse. (D) The neurotransmitter is taken back into the sending cell, a process called reuptake. (E) The neurotransmitters are either reused or broken down by monamine oxidase enzymes. Antidepressants work by interfering with steps B, D, or E, increasing the number of neurotransmitters in the synapse.

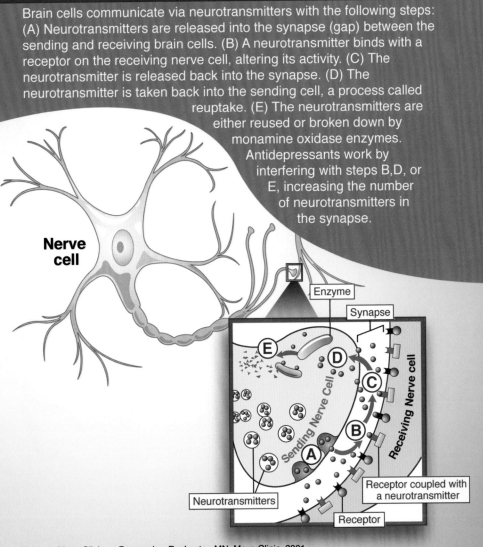

Nerve cell

Enzyme

Synapse

E · D · C · B · A

Sending Nerve Cell

Receiving Nerve cell

Receptor coupled with a neurotransmitter

Neurotransmitters

Receptor

Source: *Mayo Clinic on Depression.* Rochester, MN: Mayo Clinic, 2001.

- A 10-year study of **65,000 patients** published in the *American Journal of Psychiatry* found that the number of suicide attempts fell by **60 percent** in the first month after patients began taking antidepressants.

Global Sales of Antidepressants

Antidepressants are a multibillion dollar industry. Global sales, according to IMS Health, show that antidepressant use is growing.

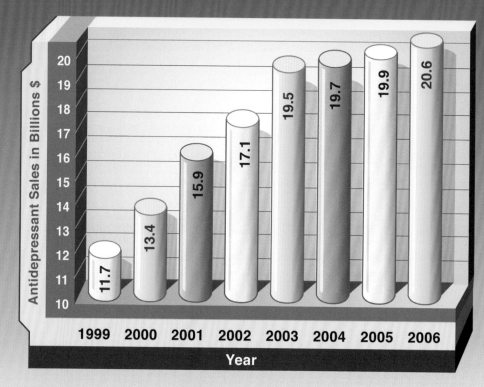

Source: IMS Health, 2007. www.imshealth.com.

- The symptoms of "**serotonin syndrome**," a result of too high levels of serotonin, include hallucinations, loss of muscular control, and confusion.

- Left untreated, **serotonin syndrome can cause death** by heart attack or circulatory failure.

- **Side effects of MAO inhibitors** include agitation, dizziness, weight changes, and gastrointestinal disturbances. It can cause dangerously high blood pressure levels if combined with certain foods.

- **Side effects of tricyclic antidepressants** (TCAs) include anxiety, difficulty concentrating, dry mouth and eyes, and appetite changes.

- Overdoses of tricyclic antidepressants can be life-threatening. TCA overdoses are the **leading cause of overdose deaths** in the United States.

- According to Thomas Moore's 1998 book *Prescription for Disaster*, the FDA had a professional staff of **1,500 to review 25 new drugs** submitted for approval, but a staff of 6 to monitor the safety of **3,000 drugs** already on the market.

Some Foods to Avoid When Taking MAO Inhibitors

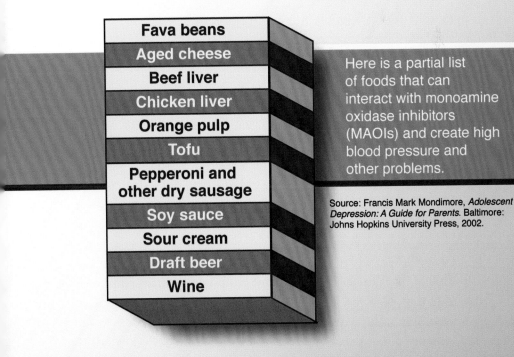

Fava beans
Aged cheese
Beef liver
Chicken liver
Orange pulp
Tofu
Pepperoni and other dry sausage
Soy sauce
Sour cream
Draft beer
Wine

Here is a partial list of foods that can interact with monoamine oxidase inhibitors (MAOIs) and create high blood pressure and other problems.

Source: Francis Mark Mondimore, *Adolescent Depression: A Guide for Parents*. Baltimore: Johns Hopkins University Press, 2002.

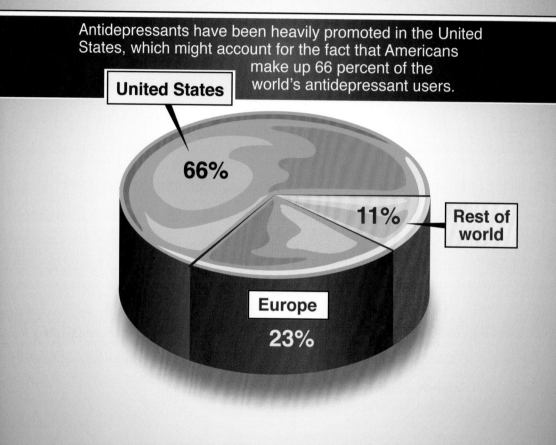

Where Consumers of Antidepressant Live

Antidepressants have been heavily promoted in the United States, which might account for the fact that Americans make up 66 percent of the world's antidepressant users.

United States

66%

11% **Rest of world**

Europe
23%

Source: *Business Wire*, December 18, 2006.

- In 1998 Donald Schell **killed his wife, daughter, and granddaughter** 3 hours after taking 2 sample capsules of Paxil; a jury in a 2001 wrongful death lawsuit ruled that Paxil's manufacturer, GlaxoSmithKline, pay $6.4 million to surviving relatives.

- A 14-month Yale University study found that 8 percent of patients admitted to the Massachusetts General Hospital psychiatric unit may have had their **mania or psychosis** induced by antidepressants.

Should Antidepressants Be Prescribed to Children and Adolescents?

> **The phenomenal growth in antidepressants prescribed to children leaves many doctors and therapists uneasy.**
>
> —Rob Waters, "A Suicide Side Effect?"

> **While most child psychiatrists agree that a minority of children may be adversely affected by antidepressants, they also believe these drugs have saved children's lives.**
>
> —E.J. Mundell, "Antidepressant Warning Might Keep Kids From Care."

A 2007 newspaper article in the *Globe and Mail* featured Sarah (not her real name), who had dropped out of high school by age 16 and was living on the streets and abusing a variety of drugs. A turning point came after she came across mindyourmind.ca, a social networking Web site that provided mental health advice to young people. Sarah determined that she was using street drugs as self-medication for her depression. "Almost three years later," writes reporter Tralee Pearce, "she is clean and sober, off the street and working full-time. She is taking Prozac to deal with depression and anxiety."[14]

Julie Woodward was 17, dealing with family conflict and romantic troubles, when she decided in 2003 to attend a local group therapy program that required its participants to take antidepressants. Her concerned parents were told that the Zoloft she was taking was both necessary and benign. But after a few days she became edgy and withdrawn. One week later she hung herself in the family garage.

As these two stories indicate, the issue of prescribing antidepressants to children and adolescents under the age of 18 often has life-or-death repercussions. If left untreated, depression can lead to ruin and possibly suicide. But concerns exist that antidepressants themselves may cause suicide among young people who take them.

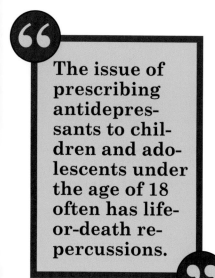

The issue of prescribing antidepressants to children and adolescents under the age of 18 often has life-or-death repercussions.

Antidepressants for Minors

Of the dozens of different antidepressants on the market, only a few have been specifically approved by the Food and Drug Administration (FDA) for use by children and adolescents. Prozac has been approved for treating depression in young people, while Prozac, Luvox, and Anafranil have been approved for treating OCD (obsessive-compulsive disorder) in people under 18.

However, many psychiatrists and pediatricians have prescribed for their young patients other antidepressants that have gotten FDA approval only for treating adults. According to a study in the *Archives of Pediatric and Adolescent Medicine*, the use of antidepressants and other psychotropic medications tripled between 1987 and 1996. An estimated 1.4 million patients under 18 are taking antidepressants, according to the *Journal of the American Academy of Child and Adolescent Psychiatry*.

One factor that disturbs some observers is that young people are being given antidepressants despite the absence of specific clinical testing for the drugs' safety and effectiveness among this population. As neuroscientist Steven E. Hyman argues, children cannot be simply viewed as miniature adults "especially because their brains are still developing." He goes on to note that there is scarce data on children who use antidepressants and that "based on existing studies, efficacy has not been clearly established for childhood depression for many antidepressants."[15] Many clinical studies have found little difference between antidepressants and a placebo. "Most children [in medical studies] get better anyway for other reasons," regardless of whether they are taking antidepressants, contends psychologist and author Michael Connor. These positive results "are often 'lumped together,' which makes the benefits of antidepressants look better than they actually are."[16]

Critics of liberally prescribing antidepressants to young people also argue that it teaches children that taking a pill is the answer to life's problems. In addition, they say, the side effects of antidepressants may interfere with a child's growth and development, and little is known about long-term health effects. Some children experience adverse reactions to antidepressants. The medical label on Luvox warns of a 4 percent risk of people developing mania or manic-like symptoms. One clinical study found a 6 percent rate of children taking Prozac developing mania—feelings of extreme joy or irritability, racing thoughts, and poor judgment and unrealistic beliefs in one's abilities. A review of the medical records of adolescent patients at Massachusetts General Hospital found that 22 percent of those treated with SSRI antidepressants suffered disturbances of mood or other adverse effects.

Antidepressants and Teen Violence

One of the adverse effects blamed on antidepressants is teen violence. In several noteworthy cases of school shootings and other acts of extreme violence, the teens involved had a history of taking antidepressants. Eric Harris, one of the 2 teens who killed 13 people and themselves in Columbine, Colorado, in 1999, was taking Luvox. Jeff Weise, 16, killed 9 persons in 2005 before committing suicide in Red Lake, Minnesota. He had a history of depression and was taking Prozac. Kip Kinkel killed his grandparents then went to school and killed 2 classmates; he had been treated with Prozac. Jason Hoffman was on the antidepressants Effexor and Celexa when he shot and wounded one teacher and 3 students at a high school in El Cajon, California. Christopher Pittman at the age of 12 shot and killed his grandparents and set their house on fire. At his trial his attorneys argued that Zoloft caused him to become violent; a jury nonetheless sentenced him to 30 years in prison in 2005.

An estimated 1.4 million patients under 18 are taking antidepressants.

Whether or not antidepressants are a cause of such acts is a matter of some controversy. Peter Breggin, a psychiatrist and longtime critic of antidepressants, contends that the known side effects of these medicines are closely linked to violent behavior. "Each of the adverse drug reactions described as 'known' antidepressants effects has been linked to violence—most obviously agitation, irritability, hostility, impulsivity, akathisia, hypomania and mania."[17] But the fact that Harris, Kinkel, and other school shooters were taking antidepressants may simply underscore their underlying mental health problems—problems that may be the true culprit behind their behavior. Although teens "may become more agitated or irritable" while on antidepressants, contends Robert Findling, chief of child and adolescent psychiatry for the University Hospitals of Cleveland, their actions are ultimately the result of their illness, not the medication. "Association is not the same as causation."[18]

The Suicide Question

Perhaps the issue that raises the most concern among parents, however, is the possibility that antidepressants may cause their children to commit suicide. Some observers such as Breggin have linked suicide to the same stimulant and mania effects they blame for violence. Other researchers speculate that antidepressants, at least at first, give depressed people enough energy and willpower to actually carry out latent suicidal impulses. As depressed people first begin drug therapy, explains psychiatrist Julio Licinio, "they experience more energy, but still feel that life isn't worth living. That's when a depressed person is most in danger of committing suicide."[19]

> " In several noteworthy cases of school shootings and other acts of extreme violence, the teens involved had a history of taking antidepressants. "

Concerns over the link between depression and suicide caused the FDA to convene a special panel and investigation in 2004. The parents of Julie Woodward were among several to testify before the panel. Most of the parents testifying also had children who attempted or succeeded in killing themselves. The FDA also reexamined 24 studies of children and teenagers with depression or other mood disorders, comparing those who took a placebo (an inactive pill) with those who took antidepressants. Of those taking the placebo, 2 out of 100 expressed suicidal thoughts or exhibited suicidal behaviors; of those on antidepressants, 4 out of 100 became suicidal. No one actually committed suicide in these study populations. But based on this analysis—and on the testimony of Julie Woodward's parents and others—the FDA decided later in 2004 to mandate that a so-called black box warning be prominently displayed on many popular antidepressants. The label warns that antidepressants are associated with "an increased risk of suicidal thinking and behavior among children and adolescents."

> "Perhaps the issue that raises the most concern among parents . . . is the possibility that antidepressants may cause their children to commit suicide."

Other countries have gone further than the United States. The British Medicines and Healthcare Products Regulatory Agency, Great Britain's equivalent to the FDA, reviewed clinical studies and directed doctors not to prescribe Paxil, Zoloft, and other popular antidepressants to children and adolescents (Prozac was permitted), arguing that the drugs' potential treatment benefits did not outweigh the risks of suicide. "These products should not be prescribed as new therapy for patients under 18 years of age with depressive illness,"[20] wrote the agency's chairman, Gordon Duff.

Did the Warning Backfire?

Although the FDA did not follow Britain's lead, its 2004 black-box warning had a significant effect. Antidepressant prescriptions for minors fell after the warning was issued in 2004. But some wonder whether this truly benefited the young people the warning was trying to help. A 2007 study in the *Journal of American Psychiatry* found that prescriptions for SSRI antidepressants fell 50 percent between 2003 and 2005—while the number of teen suicides between 2003 and 2004 jumped 18 percent. "I see parents now who are petrified to give their children antidepressants," a child psychiatrist told *New York Sun* reporter Sara Berman in 2007. "And of course I understand why. But parents of teenagers and young adults themselves need to understand that sometimes it would be scarier not to try antidepressants."[21] Simply put, the risks of leaving depression untreated can be far larger and affect greater numbers of teens. "You may induce two suicides by treatment, but by stopping treatment you're going to lose dozens to hundreds of kids," argues Robert Valuck of the University of Colorado Health Center.[22] "I think the FDA has made a very serious mistake," contends Robert Gibbon of the Center for Health Statistics. "It should lift its black-box warning because all its doing is killing kids."[23]

> Both proponents and critics of antidepressants generally agree that all children taking antidepressants should be closely monitored.

Signs to Watch For

Both proponents and critics of antidepressants generally agree that all children taking antidepressants should be closely monitored. The FDA recommends that children should see a physician at least once a week for the first month of taking antidepressants, and more often if problems arise. These problems could include

- thoughts about dying or suicide,
- new or worsening depression,
- new or worsening anxiety,
- agitation,
- panic attacks,
- insomnia,
- new or worsening irritability,
- aggressiveness,
- acting on dangerous impulses,
- unusual change in mood or behavior.

Children and teens showing one or more of these signs should consult a doctor immediately; they may end up discontinuing their antidepressants and pursuing other treatment options.

Primary Source Quotes*

Should Antidepressants Be Prescribed to Children and Adolescents?

66 At present, antidepressants are still often the first-choice treatment for young people with moderate to severe symptoms who are unable to participate fully in psychotherapy. 99

—Dwight L. Evans and Linda Wasmer Andrews," *If Your Adolescent Has Depression or Bipolar Disorder*. New York: Oxford University Press, 2005.

Evans is a professor of psychiatry, medicine, and neuroscience at the Pennsylvania School of Medicine in Philadelphia. Andrews is a science writer.

66 Like adults, children are prescribed antidepressants for a host of conditions from school phobia to attention deficit disorder to headaches. 99

—Joseph Glenmullen, *The Antidepressant Solution*. New York: Free Press, 2005.

Glenmullen is a practicing psychiatrist, a clinical instructor at Harvard Medical School, and the author of several books including *Prozac Backlash*.

Bracketed quotes indicate conflicting positions.

* Editor's Note: While the definition of a primary source can be narrowly or broadly defined, for the purposes of Compact Research, a primary source consists of: 1) results of original research presented by an organization or researcher; 2) eyewitness accounts of events, personal experience, or work experience; 3) first-person editorials offering pundits' opinions; 4) government officials presenting political plans and/or policies; 5) representatives of organizations presenting testimony or policy.

❝No radical biological shift occurs on an adolescent's eighteenth birthday that makes him suddenly an adult psychologically. Treatment decisions for some 17- and 16-year-olds may be the same as those for adults.❞

—Francis Mark Mondimore, *Adolescent Depression*. Baltimore: Johns Hopkins University Press, 2002.

Mondimore is a practicing psychiatrist and member of the clinical faculty at the Johns Hopkins University School of Medicine.

❝I've got 200 kids on an antidepressant right now, and if people took them away, I'd have parents up in arms, rioting in the streets.❞

—John Lochridge, quoted in Alyssa Abkowitz, "Have an Antidepressant Day," Creativeloafing.com, November 18, 2004.

Lochridge is a child and adolescent psychiatrist practicing in Smyrna, Georgia.

❝Antidepressants increase the risk of suicidal thinking and behavior (suicidality) in children and adolescents with major depressive disorder (MDD) and other psychiatric disorders.❞

—"Black box" warning mandated for antidepressants by the FDA in September 2004.

The Food and Drug Administration (FDA) regulates the selling and use of prescription medicines.

“The FDA's decision to implement black box warnings for all antidepressants in pediatrics has led to a precipitous decline in prescriptions. . . . I am concerned that there are many young people . . . who are not . . . receiving . . . effective treatment.”

—David Fassler, quoted in Amanda Gardner, "FDA Panel Urges Changes to Antidepressant Labeling," *Washington Post,* December 13, 2006. www.washingtonpost.com.

Fassler is a child and adolescent psychiatrist and a trustee of the American Psychiatric Association.

“Our findings strongly suggest that these individuals who committed suicide were not reacting to their [medication]. They actually killed themselves due to untreated depression.”

—Julio Licinio, quoted in Suzanne LeVert, *The Facts About Antidepressants.* New York: Marshall Cavendish, 2007.

Licinio is a psychiatry professor at the University of California at Los Angeles; he headed a study that found that less than 20 percent of suicide victims were taking antidepressants at the time of their suicide.

“There is remarkably little evidence that . . . [antidepressants] are effective in treating depression in such young patients and increasing evidence that they can lead to suicidal thoughts and behavior.”

—*New York Times,* "Risks of Antidepressants," editorial, September 16, 2004.

The *New York Times* is one of America's leading newspapers.

66We are 100 percent convinced that Zoloft killed our daughter.99

—Tom Woodward, testimony before an FDA advisory panel on antidepressants, February 2, 2004.

Woodward's daughter Julie hung herself at age 17 one week after she started taking the antidepressant Zoloft.

66Within a few weeks, I felt better. It [Lexapro] didn't change my life, but soon things seemed much more manageable. I could handle my life without so much struggling.99

—"Bill," quoted in Suzanne LeVert, *The Facts About Antidepressants*. New York: Marshall Cavendish, 2007.

Bill took Lexapro, an SSRI antidepressant, when he was 17.

66Based on the scientific evidence, it is clear that the safest and most effective treatment for childhood depression is the placebo.99

—Vera Hassner Sharav, "Tell the Truth About Antidepressants on Drug Labels and Medical Journals," Alliance for Human Research Protection, September 16, 2004.

Sharav heads the Alliance for Human Research Protection, an organization that advocates for ethical medical research practices.

66 Drug makers have been allowed to bury clinical trial results when outcomes were bad or inconclusive. That's what happened with trials of antidepressants that yielded evidence of suicidal thoughts in children. 99

—*Newsday*, "Tell Truth About Antidepressants," September 16, 2004.

Newsday is a New York–based newspaper.

Should Antidepressants Be Prescribed to Children and Adolescents?

- About **14 percent** of Americans between the ages of 12 and 17 will experience major depression at some point during their teen years.

- Every year about **500,000 children and adolescents attempt suicide**; 5,000 succeed.

- Suicide is the **third leading cause of death** for people aged 10–24.

- Doctors and psychiatrists wrote **15 million prescriptions** for antidepressants for children and adolescents in 2003.

- **Less than 5 percent** of antidepressant prescriptions in America are written for people under 18.

- SSRIs are the most common type of antidepressant prescribed for adolescents, accounting for **81 percent** of prescriptions in 2002.

- The antidepressant most commonly prescribed for children is **Zoloft**.

- Fluoxetine (Prozac) is the only antidepressant officially **approved by the FDA for adolescent use**.

- According to a 2005 report in the *Canadian Journal of Psychiatry*, **1.8 percent** of Canadian teens between the ages of 15 and 19 are taking antidepressants.

Trends in Teen Suicide

The national Youth Risk Behavior Survey (YRBS) is conducted every two years by the Centers for Disease Control and Prevention. It surveys high school students in public and private schools and seeks to monitor problems, including suicide, among America's youth.

Source: National Youth Risk Behavior Survey: 1991–2005, *Trends in the Prevalence of Suicide Ideation and Attempts*. www.cdc.gov.

- A **20 percent drop** in antidepressants prescriptions for people under 18 occurred after congressional hearings in February 2004 examined whether antidepressants caused suicide.

- Depressed adolescents who receive treatment are more likely to receive **only antidepressants rather than a combination of therapy and medication**, according to a December 2005 study published in the *Journal of Adolescent Health*.

- In clinical studies **4 percent** of young people taking antidepressants experienced suicidal thoughts or behaviors, compared with **2 percent** of those taking a placebo (dummy pill).

The Impact of the FDA's Black Box Warning on Antidepressant Prescriptions

A study of 232 physicians found that many of them decreased or stopped their prescribing of antidepressants when they learned of the FDA's warning linking antidepressant use and teen suicide. The drop was greatest among general practitioners who cared for children and adolescents.

Decreased   Stayed the same

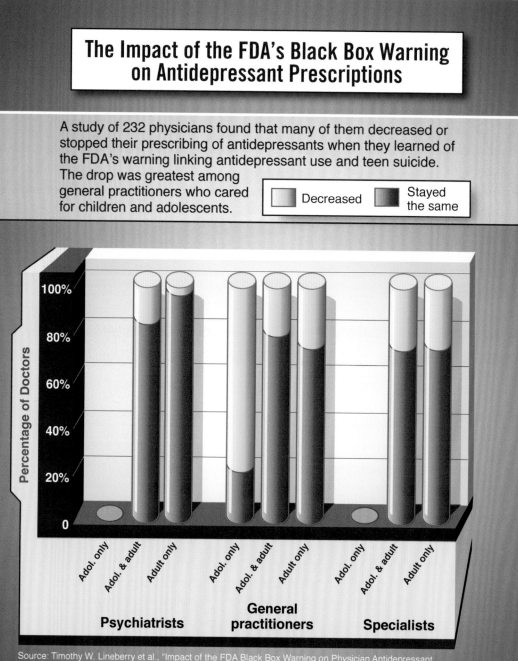

Source: Timothy W. Lineberry et al., "Impact of the FDA Black Box Warning on Physician Antidepressant Prescribing: Opening Pandora's Suicide Box," *Mayo Clinic Proceedings*, 2007.

SSRI Antidepressant Use and Suicide Rates

A study of suicide rates in 26 countries found that an increase in selective serotonin reuptake inhibitors (SSRIs), antidepressant sales in the 1990s accompanied a decline in completed suicides.

☐ Suicide rate per 100,000
☐ SSRI sales (in doses per capita)

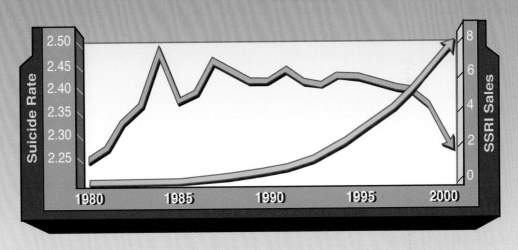

Source: Jens Ludwig et al., "Anti-Depressants and Suicide," *National Bureau of Economic Research*, 2007.

- A **22 percent drop** in antidepressant prescriptions for people under 18 occurred in the Netherlands from 2003 to 2005.

- During that same period, a **50 percent increase** in suicides among youth occurred in the Netherlands, from 34 to 51.

- America spends **more on antidepressants** for its youth than it does on antibiotics.

- In 2003 Great Britain **banned the use of all SSRI** antidepressants except fluoxetine (Prozac) for treatment of children and adolescents.

SSRIs Most Popular for Adolescents

The following bar graphs depicting the SSRI share of the total number of antidepressant prescriptions written for adolescents. Because they have fewer side effects, SSRIs are by far the most common type of antidepressants prescribed to people under 18.

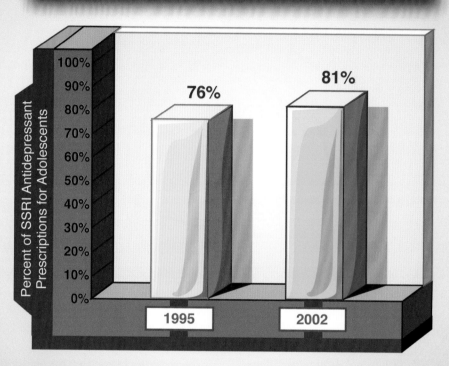

Source: *Science Daily,* November 16, 2005.

- Antidepressants are generally never prescribed to children under the age of eight.

Are Antidepressants Overprescribed?

The number of people taking antidepressant drugs increased greatly in the 1980s and 1990s when the SSRI antidepressants were introduced. The number of Americans being treated for depression jumped from 1.7 million in 1987 to 6.3 million in 1997—and the proportion of patients using antidepressants doubled from 37 percent to 75 percent. In addition, many people not clinically depressed were prescribed antidepressants for other problems or for simply feeling anxious or sad. By 2005, according to a study by the Centers for Disease Control, antidepressants were the most prescribed class of medication in the United States. Some observers worry that antidepressants, while they may be a lifesaving medication for seriously depressed people, are being overprescribed by doctors for people who do not really need them and/or may be harmed by them.

One reason touted by drug manufacturers and others for the remarkable growth in antidepressant use is the claim that Prozac and related antidepressants have fewer and less severe complications than older types of antidepressants. However, several other factors are also believed to be behind the trend of widespread antidepressant use. Among them are the decreasing social stigma of depression, the use of antidepressants for mood disorders besides depression, the popularity of prescribing antidepressants among family doctors and general practitioners, the economic incentives of America's health care system, and the marketing of antidepressants conducted by pharmaceutical companies.

> By 2005, according to a study by the Centers for Disease Control, antidepressants were the most prescribed class of medication in the United States.

New Views of Depression

In the past, depression meant loss of interest or the ability to take pleasure in any activity, to the point of being unable to sleep or eat. Such cases of major depression frequently resulted in confinement in mental health hospitals or insane asylums. Even for cases in which confinement could be avoided, a diagnosis of depression could entail years of expensive psychotherapy and perhaps taking first-generation antidepressants and experiencing their highly unpleasant side effects. Milder cases of persistent sadness not requiring such drastic intervention were often not seen as instances of depression but as simply cases of "the blues" or a personality flaw. Such people were expected to show "character" by overcoming negative emotions with willpower alone.

In recent decades the mental health profession has changed its understanding and definitions of depression. In a 1980 revision of the authoritative *Diagnostic and Statistical Manual of Mental Disorders* (DSM), the text doctors use to diagnose depression and other diseases, a distinction was made between major depression and minor depression, called dysthymia. Dysthymia has similar signs and symptoms as major depression, including difficulty in concentrating, social withdrawal, irritability, sleep problems, and weight loss or gain, but these symptoms are not as intense or as disabling as with major depression. The change in definition enabled many people who previously would not have been seen as mentally ill to be classified as depressed, enabling them to seek and receive medical treatment.

The broadening definition roughly coincided with the introduction of Prozac in 1987 and other second-generation antidepressants. As more people looked toward antidepressants to help their dysthymia, prevailing social attitudes toward depression began to change. Instead of a character flaw or weakness, depression was perceived as an illness of the brain. People who may have denied or hidden their depression because of shame now felt encouraged to seek treatment for their illness. "Once, to be depressed was to be morally and spiritually weak," says Sidney Zisook, a professor of psychiatry at the University of California at San Diego. "Now people in line at the grocery store are talking about being on Prozac. The drug has brought depression out of the closet."[24] Advocates of antidepressants argue that perhaps their greatest social contribution comes from enabling millions of people to recognize and take action against their mental illness.

> **Advocates of antidepressants argue that perhaps their greatest social contribution comes from enabling millions of people to recognize and take action against their mental illness.**

Use of Antidepressants for Other Disorders

As antidepressants became more popular in the 1990s, psychiatrists and doctors began to find other uses for them. Some of these uses have been approved by the Food and Drug Adminstration (FDA), while others have been prescribed by doctors for different problems, a practice known as "off label" prescribing. Among the mental disorders treated with antidepressants are anxiety attacks, bulimia, and obsessive-compulsive disorders.

Antidepressants also began to be tried for other, less severe conditions, such as emotional distress related to PMS or nicotine withdrawal. Some antidepressants have also been used to help some people with severe and chronic pain.

Antidepressants Prescribed by Nonpsychiatrist Doctors

Contributing to the widespread use of antidepressants was their popularity among medical doctors who previously had avoided prescribing them. Earlier antidepressants—the tricyclics and MAO inhibitors—were often very difficult to use safely or effectively. Tricyclics could be fatal in an overdose; the depressed patient taking them had to be very closely monitored by the therapist. They also had unpleasant side effects that had to be dealt with and which often caused patients to stop taking the drug. MAO inhibitors also had side effects and food interactions; a depressed patient could commit suicide by eating the wrong sort of cheese with his or her medication. As a result, most general practitioners were wary of prescribing antidepressants. Antidepressant prescriptions were mostly written by psychiatrists dealing with seriously ill people.

> **Roughly seven out of ten prescriptions for antidepressants now come from family doctors or primary care physicians.**

This changed with Prozac and other second-generation antidepressants. These drugs were not fatal in an overdose. They also had fewer and less severe side effects. They did not cause weight gain. As a result, doctors were more comfortable in prescribing them. Roughly seven out of ten prescriptions for antidepressants now come from family doctors or primary care physicians.

For many doctors, prescribing antidepressants seemed to be the appropriate response for patients who complain of general feelings of sadness, agitation, or tiredness, but who do not have any obvious physical problems. There is no blood or brain scan for depression or other mood disorders; doctors rely on questionnaires and information on the patient's emotional state. One family practitioner, Asha Wallace, prescribed antidepressants for people not necessarily depressed in clinical terms, but who "feel they're not coping, they're stressed, . . . they have too much to do just holding things together."[25] Many doctors were responding to the demands of patients themselves. This was especially true for patients whose friends had taken antidepressants and felt better. "Our phone rings off the hook every time someone does a story about Prozac," according to one doctor. "People want to try it. If you tell them they're not depressed they say, 'Sure I am!'"[26]

Economic Incentives for Antidepressants

Some aspects of America's health care system may also have contributed to the widespread use of antidepressants. Many Americans who receive health care do so through health maintenance organizations (HMOs) or other managed care institutions that limit what medical procedures are approved and paid for. Many of these organizations restrict access or set limits on the number of mental health counseling visits a patient can receive. Although many antidepressants are expensive compared to some drugs, they are often less expensive than lengthy consultations with one's doctor or weekly trips to the psychologist.

The Marketing of Antidepressants

An important factor behind the public demand for antidepressants—and the willingness of doctors to prescribe them—is the large marketing efforts made by drug companies to market and sell their product. The blockbuster success of Prozac in the late 1980s and early 1990s caused many other pharmaceutical companies to develop their own versions of antidepressants and to aggressively market them. Drug companies spend billions of dollars yearly selling antidepressants and other prescription drugs, both to doctors and to consumers.

Much of the marketing done by drug companies aimed at physicians who have the power to prescribe medications. The pharmaceutical industry spends about $13 billion annually on marketing their products to doctors—or about $13,000 per doctor. These include gifts of meals, office supplies and free samples of drugs, payment for travel to conferences and conventions, and in some cases lucrative payments or retainers to sit on advisory boards or to write articles. "Those marketing tactics are very, very effective at getting physicians to do what each drug company wants—to prescribe their product,"[27] according to David J. Rothman, a professor of social medicine at Columbia University Medical Center.

In recent years drug companies have also targeted their advertising campaigns directly at consumers. Print, television, and Internet advertising has touted the benefits of the latest antidepressants and encouraged people to talk about it with their doctors. Studies have shown that such advertising has been remarkably effective and that doctors frequently go along with prescribing medications that their patients request. Writer Margot Magowan writes that "the doctor, once in the role of diagnostician, has evolved into the drug dispenser."[28]

In some cases the marketing of antidepressants extended to defining and publicizing mood disorders themselves. One example is Glaxo-SmithKline's advertising campaign for its SSRI antidepressant Paxil. The corporation retained the services of a public relations agency to generate news stories about an illness called social anxiety disorder. The company also funded public awareness campaigns of several nonprofit organizations, including the American Psychiatric Association and the Anxiety Disorders Association of America. In part because of these publicity efforts, media accounts of and references to social anxiety disorder rose from 50 in 1997 to more than 1 billion in 1999—with most stories including the key message that Paxil was the first and only FDA-approved medication for the disorder. Sales of Paxil jumped 18 percent in 2000.

> **Print, television, and Internet advertising has touted the benefits of the latest antidepressants and encouraged people to talk about it with their doctors.**

Doctors who participated in the Paxil-sponsored public awareness campaign and other defenders of drug company advertising argue that they inform people who otherwise may be unwittingly suffering from social anxiety disorder or other disorders in ignorance, and thereby spur them to get needed medical help. But these and other marketing campaigns have raised questions about whether they are appropriate or in the public's best interest. Such ads may blur the line between a normal personality trait—such as shyness—and a severe mood disorder. They may also encourage people to take powerful drugs that are not really necessary or to eschew alternative and perhaps better treatments such as talk therapy.

Should People Work Through Sadness?

But if millions of people, perhaps encouraged by advertisement, are using antidepressants to feel less depressed or anxious, is that necessarily a bad thing? Does the fact that antidepressants are now more widely used than before mean they are being overused? Opinions on this differ, but some therapists and psychiatrists argue that antidepressants may simply mask or temporarily alleviate symptoms without solving underlying problems. They believe that antidepressants, especially SSRIs,

> Some therapists and psychiatrists argue that antidepressants may simply mask or temporarily alleviate symptoms without solving underlying problems.

are being prescribed too much to treat normal sadness—what could be seen as a normal response to divorce, job loss, house foreclosure, or other life events. In such cases antidepressants could delay the necessary healing process. People feel sad for a reason and need to examine those reasons and come to grips with them. Ronald Dworkin, a doctor and author of the book *Artificial Unhappiness: The Dark Side of the New Happy Class*, tells the story of a female patient who was upset because she wanted to take over the family finances from her husband but did not want to confront him about it. Her doctor suggested that an antidepressant would make her feel better. It did, Dworkin says, but in the end her husband's mismanagement drove the family to financial ruin. "Doctors are now medicating unhappiness," he argues. "Too many people take drugs when they really need to be making changes in their lives."[29]

Mark Linden O'Meara tried antidepressants to cope with the deaths of his parents and financial troubles. "The drugs gave me headaches and made me feel so numb." Eventually he realized that he needed to experience sadness. "I'd always worn the busy mask, denying sadness, pretending I was happy." Acknowledging and expressing his sadness enabled him eventually "to start laughing and enjoying life more."[30]

Are Antidepressants Overprescribed?

> **Antidepressants are used most often for serious depressions, but they can also be helpful for some milder depressions.**

—Mental Health America, "Medication Information," 2007. www.mentalhealthamerica.net.

Mental Health America (formally the National Mental Health Association) is a nonprofit organization dedicated to educating Americans about mental illness and helping all Americans lead mentally healthier lives.

> **Antidepressant medications do not 'cure' in the sense that antibiotics cure infections. As such, medications, even the most useful ones, are far from an ideal solution for emotional health.**

—David Servan-Schreiber, *The Instinct to Heal: Curing Stress, Anxiety, and Depression Without Drugs and Without Talk Therapy*. Emmaus, PA: Rodale, 2004.

Servan-Schreiber is a psychiatry professor at the University of Pittsburgh School of Medicine and the cofounder of the Center for Complementary Medicine at that institution.

Bracketed quotes indicate conflicting positions.

* Editor's Note: While the definition of a primary source can be narrowly or broadly defined, for the purposes of Compact Research, a primary source consists of: 1) results of original research presented by an organization or researcher; 2) eyewitness accounts of events, personal experience, or work experience; 3) first-person editorials offering pundits' opinions; 4) government officials presenting political plans and/or policies; 5) representatives of organizations presenting testimony or policy.

❝We felt it was important to encourage people with depression to speak to their doctor and not fear the stigma associated with taking prescription medication for depression.❞

—An Phan, quoted in Colby Strong, "Celebrities Join Campaign Trail Against Depression," *Neuropsychiatry*, June 2005.

Phan is a spokesperson at Pfizer, a leading pharmaceutical corporation.

❝The manufacturers are not only marketing the drug, but also the depression disease.❞

—Leslie MacKinnon, quoted in Alyssa Abkowitz, "Have an Antidepressant Day," Creativeloafing.com, November 18, 2004.

MacKinnon is a psychotherapist.

❝Seeing a psychiatrist or taking medication was not even an option in my mind. I had so many fears and misconceptions about depression and getting treatment.❞

—Lorraine Bracco, quoted in Colby Strong, "Celebrities Join Campaign Trail Against Depression," *Neuropsychiatry*, June 2005.

Bracco is an award-winning actress known for her work on *The Sopranos* television series. She participated in a marketing campaign for depression awareness sponsored by the drug company Pfizer, in which she talked about her bout with and eventual recovery from depression after seeing a psychiatrist and taking antidepressants.

❝Millions of prescriptions for SSRIs have been written. ... They are so lacking in side effects that many non-depressed people opt to take SSRIs to enhance their mood and functioning.❞

—Norman T. Berlinger, *Rescuing Your Teenager from Depression.* New York: HarperCollins, 2005.

Berlinger is a surgeon, bioethicist, and pathologist. He writes on health issues for *Discover,* the *New York Times,* and other publications.

❝Since everything else doctors do is medical, when a woman comes in who isn't feeling happy, they think antidepressants will help.❞

—John Birtchnell, quoted in "Why Millions of Women Are Hooked on the Happy Pills," *Observer,* April 18, 2004.

Birtchnell is a consultant psychiatrist with the Institute of Psychiatry in London, Great Britain.

❝For harried doctors faced with lots of patients complaining of depression, anxiety, or compulsions, drugs billed as versatile and safe can seem a godsend.❞

—David Stipp, "Trouble in Prozac," *Fortune,* November 28, 2005.

Stipp is a reporter for *Fortune,* a business newsmagazine.

66 Prozac is one of the top ten inventions of the 20th century. 99

—John Lochridge, quoted in Alyssa Abkowitz, "Have an Antidepressant Day," Creativeloafing.com, November 18, 2004.

Lochridge is a child and adolescent psychiatrist practicing in Smyrna, Georgia.

66 SSRIs are overprescribed because they were believed to be safe and the drug companies do such an incredibly effective marketing job. 99

—Peter Goldenthal, quoted in Lidia Wasowicz, "Antidepressants Often Not Effective, Healthy, Safe, or Good for Children," UPI, June 17, 2006.

Goldenthal is a family and child psychologist in Wayne, Pennsylvania.

66 [Antidepressant] ads are helpful as a tool to educate people that might need help. 99

—Darrel A. Regier, quoted in Deborah Kotz, "When Depression Goes Untreated," *U.S. News & World Report,* August 6, 2007.

Regier is the director of research for the American Psychiatric Association.

66 It's the easy solution, especially when patients demand a quick fix. 99

—Ellen McGrath, quoted in Deborah Kotz, "The Right Rx for Sadness," *U.S. News & World Report,* August 6, 2007.

McGrath is a psychologist and the author of *When Feeling Bad Is Good.*

66 Getting a prescription for an antidepressant is even easier than obtaining a flu shot. **99**

—Henry Emmons with Rachel Kranz, *The Chemistry of Joy*. New York: Fireside, 2006.

Emmons is a general and holistic psychiatrist and consultant with the Allina Medical Clinic in Northfield, Minnesota.

66 If cocaine were discovered today ... [it] would probably be promoted as a 'super neurotransmitter' reuptake inhibitor, because it boosts all three 'feel good' chemicals in the brain. **99**

—Joseph Glenmullen, *The Antidepressant Solution*. New York: Free Press, 2005.

Glenmullen is a practicing psychiatrist, a clinical instructor at Harvard Medical School, and the author of several books including *Prozac Backlash*.

Are Antidepressants Overprescribed?

- Between 1987 and 1997 the number of Americans taking antidepressants for depression grew from **600,000 to almost 5 million**.

- Almost **150 million prescriptions** were written for antidepressants in the United States in 2005.

- Antidepressants in 2005 were the **third-largest-selling** type of therapeutic drug (behind cholesterol and ulcer drugs).

- Pharmaceutical companies spend an estimated **$1.5 billion a year** to market antidepressants to doctors, according to IMS Health, a company that monitors drug sales.

- It is estimated that **one in eight Americans** has taken an antidepressant and that one in twenty are on antidepressants at any given time.

- Between **70 and 80 percent** of people with depression are treated by a primary care doctor or general practitioner, not a psychiatrist.

- **Sixty-five percent** of employee health plans limit talk therapy treatments by placing a cap on the number of mental health counseling sessions they cover.

Patient Requests Influence Doctor Prescriptions

In a study of the effects of drug advertising, researchers sent out 300 people portraying either major or minor depression to doctors. The people were further divided into those who requested antidepressants, those who requested a specific brand (Paxil), and those who did not request any medication. The study showed that patient requests had a significant impact on the doctors' prescribing practices.

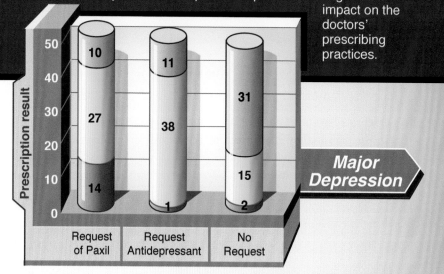

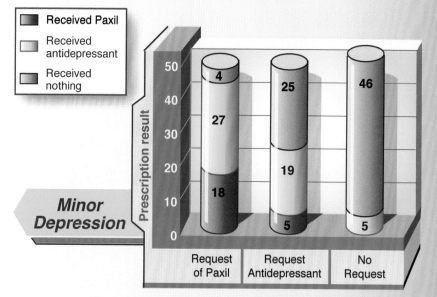

Source: Richard L. Kravitz et al., "Influence of Patients' Requests for Direct-to-Consumer Advertised Antidepressants," *JAMA*, April 27, 2005.

- People 60 and over account for **16 percent** of the U.S. population and **33 percent** of antidepressant prescriptions.

- The average cost of antidepressant treatment is **$500 a year**.

Changes in Public Opinion About Depression

Two surveys from 1991 and 2001 show how public perception of depression has shifted from viewing it as a state of mind to a physical disease.

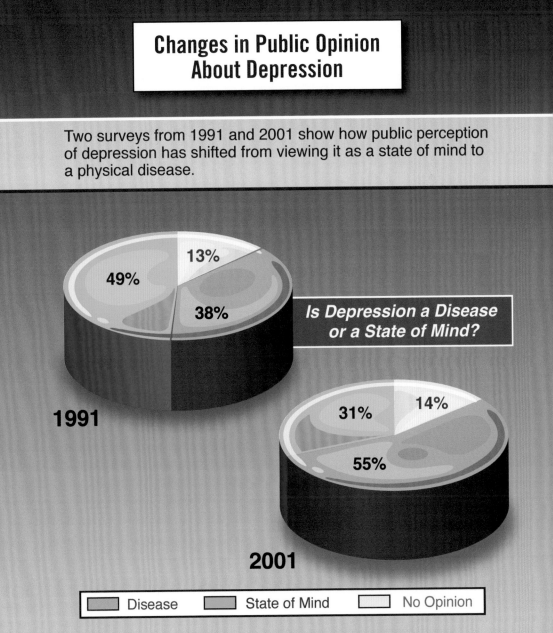

Is Depression a Disease or a State of Mind?

1991

13%

49%

38%

2001

14%

31%

55%

Disease   State of Mind   No Opinion

Source: Princeton Survey Research Associates and National Mental Health Association, 2007.

FDA-Approved Psychiatric Uses for Antidepressants

The Food and Drug Administration has approved a number of antidepressant drugs for treating a variety of mood disorders.

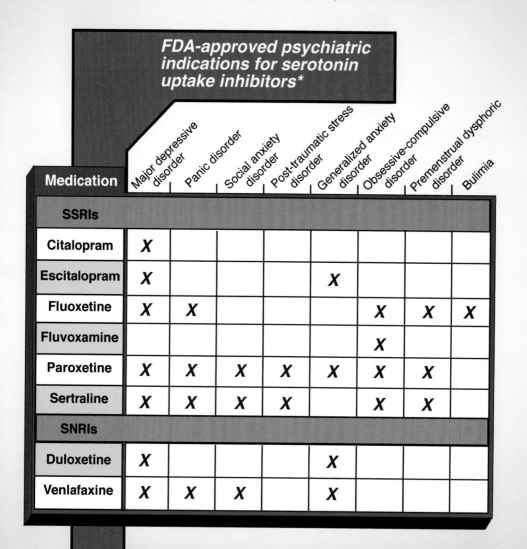

FDA-approved psychiatric indications for serotonin uptake inhibitors*

Medication	Major depressive disorder	Panic disorder	Social anxiety disorder	Post-traumatic stress disorder	Generalized anxiety disorder	Obsessive-compulsive disorder	Premenstrual dysphoric disorder	Bulimia
SSRIs								
Citalopram	X							
Escitalopram	X				X			
Fluoxetine	X	X				X	X	X
Fluvoxamine						X		
Paroxetine	X	X	X	X	X	X	X	
Sertraline	X	X	X	X		X	X	
SNRIs								
Duloxetine	X				X			
Venlafaxine	X	X	X		X			

* The absence of an X does not necessarily imply that a drug is ineffective for a given condition but, more likely, that definitive studies are lacking.

Source: James W. Jefferson, "Antidepressants: The Spectrum Beyond Depression," *Current Psychiatry,* October 2007.

Trends in Promotional Spending for Prescription Drugs

A study by the Kaiser Family Foundation examined how pharmaceutical companies were promoting their products. It showed a large increase in sending doctors free samples, significant increases in detailing (marketing directed at prescribing physicians), direct-to-consumer advertising (television and other ads targeted toward consumers), and a slight decrease in advertising in medical journals.

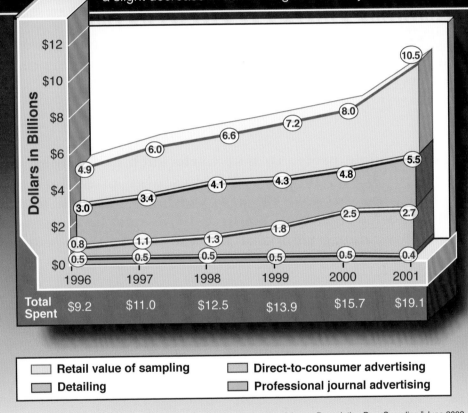

	1996	1997	1998	1999	2000	2001
Total Spent	$9.2	$11.0	$12.5	$13.9	$15.7	$19.1

Legend:
- Retail value of sampling
- Detailing
- Direct-to-consumer advertising
- Professional journal advertising

Source: Kaiser Family Foundation, "Impact of Direct-to-Consumer Advertising on Prescription Drug Spending," June 2003.

- Prozac and other antidepressants have been used to **calm down pets** and zoo animals.

Are Alternatives to Antidepressants Effective?

66 The foundation of a depression-free lifestyle consists of a good diet, a regular exercise and stress reduction program, and elimination of all chemical addictions. 99

—Hyla Cass, quoted in Randall Fitzgerald, *The Hundred Year Lie: How Food and Medicine Are Destroying Your Health.*

66 Some people prefer to take herbs such as St. John's wort or other 'natural' remedies over prescription medication because no prescription is needed and they prefer to treat themselves. This may be dangerous if the depression is severe and prevents a person from receiving proper treatment. 99

—Depression-management.info, "Herbs and 'Natural' Supplements for Depression."

Not everyone who has a depression or other mood disorder turns to antidepressants. For a variety of reasons, including costs, worries over safety, bad experiences with antidepressant side effects, or philosophical objections to taking a pill to manage their feelings, many people have pursued alternative treatments for their depression or other mood problem. These treatment options, which can either supplement or replace antidepressants, include herbal remedies such as Saint-John's-wort, diet changes and nutritional supplements, psychotherapy, and treatments that involve changes in action and lifestyle rather than medication. Whether these various options are more or less safe and effective than prescribed antidepressants is a matter of some debate.

Unlike antidepressants and other synthetic drugs, herbal remedies and nutritional supplements do not have to gain FDA approval to be produced and marketed in the United States. Manufacturers do not have to provide clinical tests demonstrating their safety or proving their claims of effectiveness (the FDA can take action against unsafe dietary supplements after they reach the market). Also, because these chemicals are not patented, companies have less economic incentive to go through the FDA process to prove their efficacy. Thus consumers are more on their own in choosing and using such products.

Herbal Remedies

So-called natural medicines are herbs and plants that have been used as drugs. They have been a part of traditional medicine in many societies, in some cases for thousands of years. Some people believe that such substances, because they are natural, are safer than synthetically produced chemicals such as Zoloft and other antidepressants. "But it's important to understand that herbal preparations are not inherently safer than or superior to chemically synthesized compounds," argues psychiatrist and author Francis Mark Mondimore, noting that "toxic substances such as nicotine and strychnine are found in plants."[31]

> " Unlike antidepressants and other synthetic drugs, herbal remedies and nutritional supplements do not have to gain FDA approval to be produced and marketed in the United States. "

Saint-John's-Wort

Saint-John's-wort (*Hypericum perforatum*) is one of the best-known herbal medicines touted for its antidepressant effects. The yellow-flowered plant herb has been used for centuries as a tea and herbal tonic for insomnia, anxiety, wound dressing, and other medical uses. Its primary compounds (hypericin and hyperforin) are available in pill form. It is the most widely prescribed antidepressant medication in Germany.

Saint-John's-wort is believed to affect the brain in ways similar to Prozac and other SSRI antidepressants by slowing down the reabsorption of the chemical serotonin, thus raising the levels of that neurotransmitter in the brain. For this reason, a person taking SSRI antidepressants should not take Saint-John's-wort at the same time.

> "Saint-John's-wort (*Hypericum perforatum*) is one of the best-known herbal medicines touted for its antidepressant effects."

Side effects of Saint-John's-wort may include dry mouth, dizziness, nausea, and increased sensitivity to sunlight. In addition, people using the herb should be very careful about drug and food interactions. If taken with an MAO inhibitor antidepressant, Saint-John's-wort could cause blood pressure to rise to dangerous levels. Taken with SSRI antidepressants such as Prozac could cause serotonin levels to rise dangerously (since Saint-John's-wort seems to raise serotonin levels as well). Saint-John's-wort may also reduce the effectiveness of some anticancer drugs and oral contraceptives.

Scientific studies of Saint-John's-wort's effectiveness have had mixed results. A systematic review of 23 clinical studies published in the *British Medical Journal* in 1996 concluded that extracts of Saint-John's-wort "are more effective than placebo for the treatment of mild to moderately severe depressive disorders"[32] and that patients experienced fewer and less severe side effects than those taking standard antidepressants. But the authors of an American clinical trial published in the April 18, 2001, issue of the *Journal of the American Medical Association* found no "significant antidepressant or anti-anxiety effects for St. John's wort when compared to placebo" and that the herb was "not effective for treatment of major depression."[33]

Ginkgo Biloba

The ginkgo biloba tree has been traced back 200 million years; it is believed to be the oldest living species of tree. Its seeds and leaves have been ingredients in Chinese medicine for thousands of years; it is also highly popular in Europe.

The primary effect of ginkgo is to improve blood flow to the brain and other parts of the body. Some studies have indicated that this may improve mood, especially in elderly people. Some doctors prescribe gingko to counteract the sexual dysfunction side effects of standard antidepressants. Ginkgo can interfere with blood clotting and should be avoided by people with bleeding problems.

Siberian Ginseng

Siberian ginseng (*Eleutherococcus senticosus*) is a shrub indigenous to the mountainous regions of east Asia. It has long been used in Chinese medicine as a way to combat insomnia and restore energy. It has been known to elevate the mood of depressed people, and some animal studies have indicated that it acts similarly to prescription MAO inhibitors. Further clinical studies are needed to demonstrate its effectiveness on people with depression.

Nutritional Supplements

In recent years much research and interest has focused on how eating habits and nutrition affect the cells and neurotransmitters of the brain. Some psychiatrists assess what their patients eat and prescribe nutritional supplements and changes in diet instead of antidepressants. In addition to making sure people get enough of such nutrients as vitamins B6 and B12, folic acid, magnesium, iron, and zinc, these therapists often suggest lesser-known nutritional supplements.

S-adenosylmethionine, or SAM-e, is an amino acid naturally occurring in all living cells. Several European studies have suggested that it can work as a mild antidepressant with fewer side effects. It is believed to help brain cells be more responsive to neurotransmitters such as serotonin and dopamine. It is used in Europe as a prescription drug. In the United States it has been available as a nutritional over-the-counter supplement since 1999 but has not been approved by the FDA for treating depression. Side effects include nausea and constipation.

> " Some psychiatrists assess what their patients eat and prescribe nutritional supplements and changes in diet instead of antidepressants. "

The amino acid 5-hydroxytryptophan, or 5-HTP, is converted in the brain into the neurotransmitter serotonin. Thus it increases the serotonin level in the brain, which may alleviate depression symptoms. It is prescribed in Europe to treat depression and is available in the United States as a nutritional supplement. Studies have suggested that it may relieve depression in some patients, although the results are not conclusive. "Larger studies than have been conducted to date are needed,"[34] wrote *Wall Street Journal* reporter Nancy Keates in July 2007.

Omega-3 fatty acids are found in fish oils and certain plants, including flax seed. Fish are a leading source of such fats, especially salmon, mackerel, tuna, and other fatty fish. Omega-3 can also be taken as fish oil tablets or other nutrition supplements. Mega-3s have been touted by some as an effective treatment for depression. A letter published in the *Archives of General Psychiatry* in January 2002 describes a 21-year-old who had been battling depression for seven years and who had become actively suicidal. Antidepressants such as paroxetine (Paxil) had failed to help. He was given eicosapentaenoic acid (EPA), one of the omega-3 fatty acids. His suicidal thoughts quickly ceased, and within nine months all of his depressive symptoms were gone. "As the only change in therapy during the 9-month period was the addition of EPA, . . . it seems likely that the clinical improvement was associated with the EPA,"[35] concluded the letter's authors, who called for more studies of the substance. Several studies have been made, including one in Israel that seemed to find therapeutic benefits for children suffering from depression, but taken as a whole the studies "don't provide enough evidence to provide a clear conclusion,"[36] according to Keates.

Talk Therapy

Talk therapy, or psychotherapy, involves talking to a trained professional about one's feelings, problems, and behaviors. It can be done one-on-one with a psychologist or other therapist, with family members, or in small groups. "While antidepressants address the chemical bases of depression, psychotherapy targets the psychological, social, and behavioral aspects of the illness," writes psychiatrist Dwight L. Evans. "Interestingly, recent research using sophisticated brain imaging technology shows that psychotherapy may lead to physical changes in brain pathways as well. However, the pattern of brain changes is somewhat different from that seen in people taking antidepressants."[37]

Advocates of talk therapy argue that it can create lasting cures for depression without the physical side effects of antidepressants. Some clinical studies have found that some forms of talk therapy are as much or more effective than antidepressants. Most health care experts agree that the best treatment for depression and other mood disorders usually involves a combination of medication and talk therapy.

As with antidepressants, different types of talk therapy are used. Often therapists combine elements of different therapies. Two of the most popular types for depression are cognitive-behavioral therapy and interpersonal therapy.

Cognitive-behavioral therapy (CBT) helps people identify and isolate the patterns and habits of negative thoughts and feelings about themselves and their environment, and to work to change these negative thoughts (and behaviors) into positive ones. One technique is to teach people to break down large tasks into smaller and more manageable parts; another is to practice social skills and coping strategies to better deal with situations. Patients are taught to monitor their moods, manage stress, and set and achieve goals.

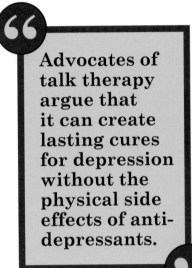

> Advocates of talk therapy argue that it can create lasting cures for depression without the physical side effects of antidepressants.

Interpersonal therapy (IPT) focuses on interpersonal or relationship issues that may have triggered the depression, such as a divorce, parental conflict, or problems with friends. Patients in IPT learn to understand people, to express their feelings more clearly, and to develop social and communication skills to solve relationship problems.

Electroconvulsive Therapy

Another, somewhat more controversial, treatment for depression is electroconvulsive therapy (ECT). Developed in the 1930s, by the 1950s ECT or shock therapy had become the standard treatment for severe depression. It has developed a scary reputation partially because of media portrayals (such as in the book and movie *One Flew Over the Cuckoo's Nest*), and partially because in its early years it *was* scary; a fully awake patient would be forcibly held down by hospital staff while a strong electrical current was used to trigger a seizure. However, ECT in modern times is much different. Patients are given a muscle relaxant and an anesthetic, while small electrodes are placed on the head. Then small precise amounts of electric current are sent to the brain to trigger a brief seizure. This is repeated several times per treatment, which is generally given three times a week.

How exactly ECT works to alleviate depression symptoms is unknown; some people have compared it to rebooting a computer system to get the electrochemical processes of the brain functioning normally. Side effects include headaches, soreness, nausea, and memory loss. It is generally used only on patients with severe depression or who have not responded to antidepressants or other treatments.

The idea of applying electrical stimulation to the brain is behind some newer experimental treatments of depression. Vagus nerve stimulation (VNS) involves a small battery-powered device that is surgically implanted in a patient's chest. Much like a pacemaker, the device delivers mild electric shocks to the vagus nerve, a nerve that is connected to the parts of the brain that regulate mood. Originally developed to treat epilepsy, VNS has shown promise in treating depression.

Other Alternatives to Antidepressants

Many of the suggestions for alleviating depression without taking antidepressants are activities that promote general physical and mental health. Getting enough sleep is one way to cope with stress, as is getting enough exercise. As little as 30 minutes a day of active exercise can alleviate chronic sadness as much as antidepressants, according to a 2005 study in the *American Journal of Preventive Medicine*. Robert Thayer, a professor of psychology at California State University, argues that even a simple 15-minute walk "can improve your mood and increase your energy for up to two hours."[38] In addition to aerobic exercises, some mental health practitioners recommend yoga or similar breathing exercises to reduce breathing rate, heart rate, and tension. Other suggested activities for people combating feelings of anxiety or depression include listening to music, joining a support group, or getting involved in community or volunteer work. "The best remedies for mild sadness? Happy actions, not happy pills,"[39] writes medical journalist Deborah Kotz.

> **Many of the suggestions for alleviating depression without taking antidepressants are activities that promote general physical and mental health.**

Are Alternatives to Antidepressants Effective?

❝St. John's wort is an herb that has become extremely popular for the over-the-counter treatment of depression, even though there is no reliable scientific evidence that it is effective.❞

—Harold S. Koplewicz, *More than Moody: Recognizing and Treating Adolescent Depression.* New York: G.P. Putnam's Sons, 2002.

Koplewicz is a child psychiatrist and founder of the New York University Child Study Center.

❝This research suggests that major depression of at least moderate severity should not be treated with St. John's wort.❞

—Jonathan Davidson, quoted in Debra Goldschmidt, "Study Questions St. John's Wort's Effectiveness," CNN, April 10, 2002. www.cnn.com.

Davidson was the lead researcher in a 2002 Duke University Medical Center clinical trial—the largest ever at that time—of Saint-John's-wort.

* Editor's Note: While the definition of a primary source can be narrowly or broadly defined, for the purposes of Compact Research, a primary source consists of: 1) results of original research presented by an organization or researcher; 2) eyewitness accounts of events, personal experience, or work experience; 3) first-person editorials offering pundits' opinions; 4) government officials presenting political plans and/or policies; 5) representatives of organizations presenting testimony or policy.

66Instead of carefully weighing the risks of antidepressants against the benefits, [some] . . . groups push the agenda that mental illness is a controllable state of mind, not a disease.**99**

—Darrel A. Regier, quoted in Deborah Kotz, "When Depression Goes Untreated," *U.S. News & World Report,* August 6, 2007.

Regier is director of research at the American Psychiatry Association.

66Taking 5-HTP may have an even more beneficial effect than antidepressants because instead of merely preventing the recycling of serotonin, 5-HTP actually increases serotonin levels.**99**

—Henry Emmons with Rachel Kranz, *The Chemistry of Joy.* New York: Fireside, 2006.

Emmons practices general and holistic psychiatry and has conducted workshops for health care professionals on alternative and natural therapies.

66Although ECT is effective, it causes pronounced memory problems and its effects are transitory.**99**

—Edward Drummond, *The Complete Guide to Psychiatric Drugs.* Hoboken, NJ: John Wiley & Sons, 2006.

Drummond is associate medical director at the Seacoast Mental Health Center in Portsmouth, New Hampshire.

❝ECT has gotten me off antidepressants. . . . After ECT, I was able to work on issues that I couldn't before.❞

—Kitty Dukakis with Larry Tye, "'I Feel Good, I Feel Alive,'—in a New Book, Kitty Dukakis Credits Electroconvulsive Therapy for Relieving Her Famously Disabling Depression," *Newsweek,* September 18, 2006.

Dukakis, the wife of former Massachusetts governor and presidential candidate Michael Dukakis, detailed her longtime battle with depression and her experience with ECT in the book *Shock,* cowritten with Larry Tye.

❝Omega-3s help regulate mental-health problems because they enhance the ability of brain-cell receptors to comprehend mood-related signals from other parts of the brain.❞

—David B. Wexler, *Is He Depressed or What?* Oakland, CA: New Harbinger, 2005.

Wexler is a clinical psychologist in San Diego specializing in the treatment of people with relationship conflicts.

❝A well-balanced nutrition program can serve as a strong foundation for healing depression.❞

—Lewis Harrison, *Healing Depression Naturally.* New York: Kensington, 2004.

Harrison is an instructor at the New York Botanical Garden and the author of several books on herbal medicine, nutrition, and natural healing.

"Although cutting-edge research in nutritional neuro-science holds much promise, the evidence does not yet strongly support changes in diet alone as an effective way to deal with depression."

—John Preston, *You Can Beat Depression: A Guide to Prevention and Recovery,* 4th ed. Atascadero, CA: Impact, 2004.

Preston is a professor of psychology at Alliant International University and the author of 14 books.

"Neither medication nor therapy is particularly more effective than the other; for me personally it had to be both."

—Anthony Sampson, quoted in Ava T. Albrecht and Charles Herrick, *100 Questions & Answers About Depression.* Sudbury, MA: Jones and Bartlett, 2006.

Sampson is a foreign language teacher and cancer survivor who has received on-going care from a psychiatrist to manage his depression and anxiety.

"The problem per se is not that there are side effects; it is that herbal treatments are not regulated as to either their safety or efficacy."

—Ava T. Albrecht and Charles Herrick, *100 Questions & Answers About Depression.* Sudbury, MA: Jones and Bartlett, 2006.

Albrecht is a practicing psychiatrist in New York City and a member of the faculty at the New York University School of Medicine. Herrick is medical director for Intensive Psychiatric Services at Danbury Hospital in Danbury, Connecticut, and a faculty member at New York Medical College.

66 The effects of exercise therapy appear to be unusually durable, typically more long lasting than the effects of psychotherapy and drug treatment.99

—Keith Johnsgard, *Conquering Depression & Anxiety Through Exercise*, Amherst, NY: Prometheus, 2004.

Johnsgard is a retired professor of psychology at San Jose State University, a distance runner, and the author of several books and articles on exercise and depression.

66 Doing something nice for someone else is one of the best ways we can think of to extricate yourself from a bad mood.99

—Laura L. Smith and Charles H. Elliot, *Depression for Dummies.* Indianapolis: Wiley, 2003.

Smith and Elliott are clinical psychologists and specialists in cognitive therapy.

Are Alternatives to Antidepressants Effective?

- **Saint-John's-wort (*Hypericum perforatum*)** is an herb that has been used for centuries to calm anxiety and as a balm for wounds.

- In Germany doctors prescribe or recommend Saint-John's-wort about **20 times more often** than Prozac.

- The **active ingredient(s)** in Saint-John's-wort remain an object of scientific research. Some scientists posit that the substance hyperforin accounts for the herb's antidepressant effects.

- Reported **side effects of Saint-John's-wort** include dry mouth, nausea, dizziness, fatigue, and increased sensitivity to sunlight.

- Herbal medicines and dietary supplements are under **less FDA scrutiny** and regulation than prescription drugs.

- The exact ingredients and **manufacturing quality** of herbs and nutritional supplements **may vary** from manufacturer to manufacturer.

- **SAM-e has been available** in the United States since 1999.

- **SAM-e has not been approved** by the FDA for use as an antidepressant.

- People using **herbal remedies or nutritional supplements** should tell their doctor what they are taking.

Key Studies of Saint-John's-Wort for Depression

Saint-John's-wort is a flowering plant that has been widely studied for its medical uses. This table summarizes some of the clinical studies of St. John's wort for treating depression.

Study/location	Agents/dosage	Outcome
Cochrane Review Linde, 1996 various locations	Saint-John's-wort (350 to 1,800 mg) daily	Saint-John's-wort was superior to placebo and as effective as standard antidepressants.
Wheatley, 1997 United Kingdom	Saint-John's-wort (900 mg) versus amitriptyline (Elavil; 75 mg) daily for 6 weeks	Both treatments were equally effective.
Philipp, 1999 Germany	Saint-John's-wort (1,050 mg) versus imipramine (Tofranil; 100 mg) versus placebo daily	Saint-John's-wort was more effective than placebo and as effective as imipramine.
Harrer, 1999 Austria	Saint-John's-wort (800 mg) versus fluoxetine (Prozac; 20 mg) daily for 6 weeks	Both treatments were equally effective.
Schrader, 2000 Germany	Saint-John's-wort (500 mg) versus fluoxetine (20 mg) daily for 6 weeks	Both treatments were equally effective.
Woelk, 2000 Germany	Saint-John's-wort (500 mg) versus imipramine (150 mg) daily for 6 weeks	Both treatments were equally effective.
Brenner, 2000 United States	Saint-John's-wort (900 mg) versus sertraline (Zoloft; 75 mg) daily for 6 weeks	Saint-John's-wort was at least as effective as sertraline.
Kalb, 2001 Germany	Saint-John's-wort (900 mg) versus placebo daily for 42 days	Saint-John's-wort was superior to placebo at days 28 and 42.
Vorbach, 1997 multicenter	Saint-John's-wort (1,800 mg) versus imipramine (150 mg) daily for 6 weeks	Both treatments were equally effective (HAM-D).
Shelton, 2001 United States	Saint-John's-wort (900 mg, increased to 1,200 mg if needed) versus placebo daily for 4 weeks	Proportion achieving response did not differ between groups.
Hypericum Depression Trial Study Group, 2002 multicenter	Saint-John's-wort (900 to 1,500 mg) versus sertraline (50 to 100 mg) versus placebo daily for 8 weeks	Neither sertraline nor Saint-John's-wort was significantly different from placebo.

Source: Silvana Lawvere and Martin C. Mahoney, "St. John's Wort," *American Family Physician*, December 1, 2005.

- Exercise causes the brain to **produce endorphins**, natural pleasure-creating chemicals.

- As many as **100,000 people** each year receive electroconvulsive therapy (ECT) in the United States.

What Licensed Mental Health Professionals Are Allowed to Do

Many mental health providers can provide therapy but cannot prescribe medication.

Health care professional	May prescribe medication	May provide psychotherapy
Psychiatrists	Yes	Yes
Primary care doctors	Yes	No
Psychiatric nurses	Yes (with advanced training)	Yes
Clinical psychologists	No	Yes
Clinical social workers	No	Yes
Mental health counselors	No	Yes
Marriage/family therapists	No	Yes

Note: Two states (New Mexico and Louisiana) have recently passed laws granting prescribing privileges to some psychologists.

Source: Dwight L. Evans and Linda Wasmer Andrews, *If Your Adolescent Has Depression or Bipolar Disorder.* New York: Oxford University Press, 2005.

Omega-3 Fatty Acids in Fish

Omega-3 Fatty Acids, including Eicosapentaenoic Acid (EPA) and Docosahexaenoic Acid (DHA), are a class of nutrients believed to be helpful in preventing and/or treating depression. Some kinds of fish are rich sources of these nutrients.

Food	Serving	EPA (g)	DHA (g)	Amount providing 1 g of EPA + DHA
Herring, Pacific	3 oz	1.06	0.75	1.5 oz
Salmon, chinook	3 oz	0.86	0.62	2 oz
Sardines, Pacific	3 oz	0.45	0.74	2.5 oz
Salmon, Atlantic	3 oz	0.28	0.95	2.5 oz
Oysters, Pacific	3 oz	0.75	0.43	2.5 oz
Salmon, sockeye	3 oz	0.45	0.60	3 oz
Trout, rainbow	3 oz	0.40	0.44	3.5 oz
Tuna, canned, white	3 oz	0.20	0.54	4 oz
Crab, Dungeness	3 oz	0.24	0.10	9 oz
Tuna, canned, light	3 oz	0.04	0.19	12 oz

Source: "Essential Fatty Acids," The Linus Pauling Institute Micronutrient Information Center, December 7, 2005.

- Eliminating **alcohol, caffeine, nicotine, and illegal drug** use can help a person's depression.

- All approaches to depression treatment take some **effort on the part of the patient**.

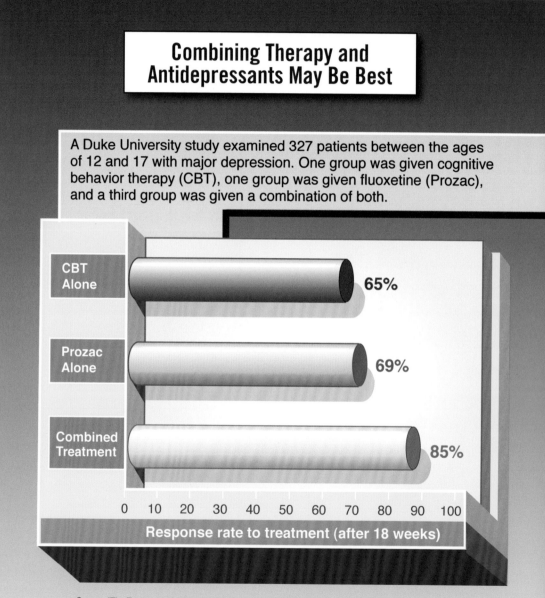

Combining Therapy and Antidepressants May Be Best

A Duke University study examined 327 patients between the ages of 12 and 17 with major depression. One group was given cognitive behavior therapy (CBT), one group was given fluoxetine (Prozac), and a third group was given a combination of both.

CBT Alone — 65%

Prozac Alone — 69%

Combined Treatment — 85%

0 10 20 30 40 50 60 70 80 90 100

Response rate to treatment (after 18 weeks)

Source: "The Treatment of Adolescents with Depression Study," *Archives of General Psychiatry*, October 2007.

Vagus Nerve Stimulation (VNS)

The Vagus nerve is crtical to the body, connecting the brain stem with the heart, lungs, and intestines. Vagus nerve stimulation (VNS) consists of a surgically inserted electrical pulse generator that stimulates the vagus nerve. In 2005 the FDA approved VNS as a treatment for depression that is resistant to both medication and therapy.

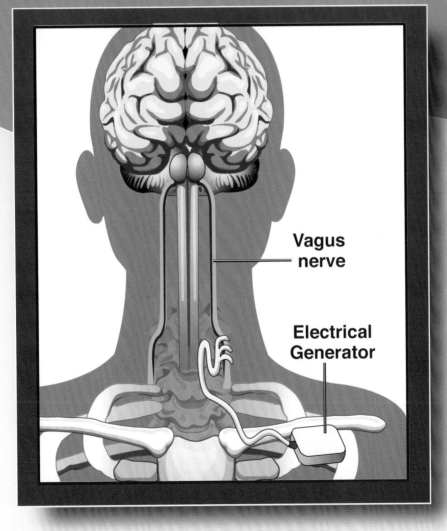

Vagus nerve

Electrical Generator

Source: *Mayo Clinic on Depression*. Rochester, MN: Mayo Clinic, 2001.

Key People and Advocacy Groups

Peter Breggin: Breggin is a psychiatrist and author who has written several books highly critical of antidepressants and other medicines for mental illnesses.

Arvid Carlsson: Carlsson, a Swedish scientist, won the 2000 Nobel Prize in Medicine for his work on how the brain functions, including the discovery that dopamine was a neurotransmitter. His research helped pave the way for the development of SSRI antidepressants.

Hyla Cass: Cass is a psychiatrist and author of many books that expound on the idea that depression can and should be treated with diet changes and nutritional supplements rather than antidepressants.

Citizen's Commission on Human Rights: CCHR is a leading advocacy group that strongly opposes the use of antidepressants (and other psychiatric drugs). It was founded in 1969 by the Church of Scientology.

Eli Lilly and Company: Eli Lilly is an Indianapolis-based drug manufacturer; its multibillion-dollar success with Prozac in the 1980s and 1990s transformed both the corporation and the psychiatric drug industry.

Ray Fuller, David Wong, and Byran Molloy: Fuller, Wong, and Molloy are research scientists at Eli Lilly who are credited with developing the chemical compound that became known as Prozac.

GlaxoSmithKline: Formed by the merger of SmithKline Beecham and Glaxo Wellcome, GlaxoSmithKline is the manufacturer of the antidepressants Paxil and Wellbutrin.

Joseph Glenmullin: Glenmullin is an author and psychiatrist who has argued that antidepressants cause chemical dependency and have long-term risks that will be revealed over time.

David Healy: Healy is an Irish psychiatrist and a professor of psychological medicine at the University College of Medicine in Cardiff, Wales. He has been an expert witness in several lawsuits brought against antidepressant manufacturers, arguing that they have hidden the risks of antidepressants to the public.

Nathan S. Kline: Kline was a psychiatrist and one of the leading discoverers and promoters of iproniazid (trade name Marsalid) in the 1950s. Iproniazid, an MAO inhibitor, was one of the first drugs to be used as an antidepressant.

Peter D. Kramer: Kramer is a psychiatrist and author of *Listening to Prozac* (1993), a best-selling study of that antidepressant. Kramer speculated that it made his patients "better than well" and coined the term *cosmetic psychopharmocology* to describe the use of medicine in nondepressed people for the sole purpose of creating wanted personality traits.

Ronald Kuhn: Kuhn was a Swiss psychiatrist and pharmacologist who developed imipramine, the first of the tricyclic antidepressants, in the 1950s.

Pfizer Corporation: Pfizer is a large pharmaceutical corporation whose leading sellers include the antidepressant Zoloft.

Brooke Shields and Tom Cruise: Shields and Cruise are celebrity movie stars who got into a public debate over whether mothers suffering from postpartum depression should take antidepressants.

Andy Vickery: Vickery, a Houston-based lawyer, has been the lead attorney in several civil lawsuits against the manufacturers of antidepressants on behalf of people who claimed the drugs caused suicide, homicide, and acts of violence.

Chronology

1958
Iproniazid (trade name Marsalid), an MAO inhibitor antidepressant, is first marketed in the United States.

1906
Congress passes the federal Food and Drug Act to ensure the safety of foods and medicines and creates the Food and Drug Administration.

1962
Congress gives the FDA power to withhold drugs from the market unless scientific studies can attest to their safety and efficacy for specific conditions.

1948
American scientists isolate the neurotransmitter serotonin.

1984
The German government approves the use of Saint-John's-wort for depression.

| 1905 | 1925 | 1945 | 1965 | 1985 |

1938
The federal Food, Drug, and Cosmetic Act requires that drugs be safe before they are put on the market.

1987
Fluoxetine is approved for depression treatment by the FDA and is marketed under the trade name Prozac.

1951
The Humphrey-Durham Amendment to the 1938 Food, Drug, and Cosmetic Act makes psychotropic drugs, including antidepressants, available by prescription only.

1972
Fluoxetine is developed by researchers at Eli Lilly and Company.

1960
Imipramine (trade name Tofranil) is introduced to the United States; it is the first of the tricyclic antidepressants.

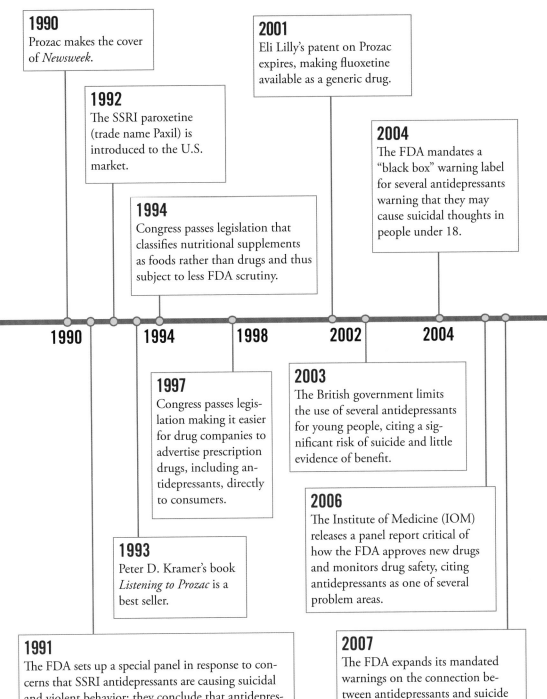

1990
Prozac makes the cover of *Newsweek*.

2001
Eli Lilly's patent on Prozac expires, making fluoxetine available as a generic drug.

1992
The SSRI paroxetine (trade name Paxil) is introduced to the U.S. market.

2004
The FDA mandates a "black box" warning label for several antidepressants warning that they may cause suicidal thoughts in people under 18.

1994
Congress passes legislation that classifies nutritional supplements as foods rather than drugs and thus subject to less FDA scrutiny.

1990   1994   1998   2002   2004

1997
Congress passes legislation making it easier for drug companies to advertise prescription drugs, including antidepressants, directly to consumers.

2003
The British government limits the use of several antidepressants for young people, citing a significant risk of suicide and little evidence of benefit.

2006
The Institute of Medicine (IOM) releases a panel report critical of how the FDA approves new drugs and monitors drug safety, citing antidepressants as one of several problem areas.

1993
Peter D. Kramer's book *Listening to Prozac* is a best seller.

1991
The FDA sets up a special panel in response to concerns that SSRI antidepressants are causing suicidal and violent behavior; they conclude that antidepressants are not to blame. The SSRI sertraline (trade name Zoloft) is introduced to the U.S. market.

2007
The FDA expands its mandated warnings on the connection between antidepressants and suicide thoughts to encompass young adults aged 18–24.

Related Organizations

Alliance for Human Research Protection (AHRP)

142 West End Ave., Suite 28P

New York, NY 10023

Web site: www.ahrp.org

AHRP is a nonprofit advocacy group that works to protect human volunteers in medical clinical trials. It has published numerous articles critical of how antidepressants have been tested and marketed and advocates full disclosure of all possible side effects of antidepressants. Much of its published work is available on its Web site.

Alternative Medicine Foundation (AMF)

PO Box 60016

Potomac, MD 20859

phone: (301) 340-1960

fax: (301) 340-1936

Web site: www.amfoundation.org

The Alternative Medicine Foundation is a nonprofit organization that seeks to provide responsible and reliable information about alternative medicine to the public and health professionals. Its Web site features a searchable guide on herbal medicines such as Saint-John's-wort.

American Academy of Child & Adolescent Psychiatry

3615 Wisconsin Ave. NW

Washington, DC 20016

phone: (800) 333-7636

fax: (202) 966-2891

Web site: www.aacap.org

The AACAP is a national organization of medical professionals who research, evaluate, and treat children with depression and other mental, behavioral, and developmental disorders. It publishes the monthly *Journal of the American Academy of Child and Adolescent Psychiatry*. Fact sheets and other informational materials are available on its Web site.

American Foundation for Suicide Prevention (AFSP)

120 Wall St., 22nd Floor

New York, NY 10005

phone: (888) 237-2377

fax: (212) 363-3500

e-mail: inquiry@afsp.org

Web site: www.afsp.org

The AFSP supports scientific research on suicide, including research on depression and its treatment. It works to educate the public on the recognition and treatment of depressed and suicidal individuals. It publishes the newsletter *Crisis*.

American Psychological Association (APA)

750 First St. NE

Washington, DC 20002-4242

phone: (202) 336-5500

fax: (202) 336-5708

e-mail: public.affairs@apa.org

Web site: www.apa.org

The APA is the world's largest organization of psychologists. It publishes numerous books, journals, and videos and has information on depression treatment and psychotherapy on its Web site.

Citizen's Commission on Human Rights (CCHR)

6616 Sunset Blvd.

Los Angeles, CA 90028

phone: (800) 869-2247

fax: (323) 467-3720

e-mail: humanrights@cchr.org

Web site: www.cchr.org

CCHR is a nonprofit organization whose goal is to expose what it considers to be human rights abuses by psychiatry. The organization believes that drugs prescribed by psychiatrists, including antidepressants, can cause violence and serious mental health problems. CCHR is sponsored by the Church of Scientology.

Depression and Bipolar Support Alliance (DBSA)

730 N. Franklin St., Suite 501

Chicago, IL 60610

phone: (800) 826-3632

fax: (312) 642-7243

Web site: www.dbsalliance.org

DBSA is a patient-directed national organization that works to educate the public about mood disorders and to ensure that people with disorders are treated equitably. It distributes free brochures and other educational materials on mood disorders and treatments.

Mental Health America (MHA)

2000 N. Beauregard St., 6th Floor

Alexandria, VA 22311

phone: (703) 684-7722

fax: (703) 684-5968

toll free: (800) 969-6642

e-mail: nmhainfo@aol.com

Web site: www.nmha.org

www.mentalhealth.net

Mental Health America, formally the National Mental Health Association, is a national nonprofit organization that works to promote good mental health for all Americans. Information on mental disorders and medications is available on its Web site. It also publishes books and pamphlets on overcoming mental illness.

National Alliance for the Mentally Ill (NAMI)

Colonial Place Three, 2107 Wilson Blvd., Suite 300

Arlington, VA 22201

phone: (703) 524-7600

fax: (703) 524-9094

Web site: www.nami.org

NAMI is a support and advocacy organization of people with severe mental illness and their friends and families. It has more than 1,200 local affiliates. Its Web site has information on depression, antidepressants and other treatments, and places for finding support.

National Center for Complementary and Alternative Medicine (NCCAM)

NCCAM Clearinghouse, P.O. Box 7923

Gaithersburg, MD 20898

phone: (888) 644-6226

fax: (866) 464-3616

e-mail: info@nccam.nih.gov

Web site: http://nccam.nih.gov

NCCAM, part of the National Institutes of Health (NIH), is the lead agency of the American federal government for scientific research on complementary and alternative medicine—medicine that is outside of conventional medical practices. Its Web site includes fact sheets and links on alternative medicine, including information on herbal antidepressants and alternative treatments for depression.

National Institute of Mental Health, Office of Communications

6001 Executive Blvd., Room 8814, MSC 9663

Bethesda, MD, 20892-9663

phone: (866) 615-6464

fax: (301) 443-4279

e-mail: nimhinfo@nih.gov

Web site: www.nimh.nih.gov

NIMH is the federal agency concerned with mental health research and education. It funds research on the causes and treatment of depression and other mental diseases and produces various informational publications on mental disorders and their treatment.

The Royal College of Psychiatrists

17 Belgrave Sq.,

London SW1X 8PG, Great Britain

phone: (020) 7253-2351

fax: (020) 7245-1231

e-mail: rcpsych@rcpsych.ac.uk

Web site: www.rcpsych.ac.uk

The Royal College is the educational and professional body for psychiatrists in Great Britain and Ireland. It publishes several journals, including *Advances in Psychiatric Treatment*. It also publishes several pamphlets and articles on antidepressants, alternative treatments for depression, and other areas on its Web site.

United States Food and Drug Administration (FDA)

5600 Fishers Ln.

Rockville, MD 20857-0001

phone: (888) 463-6332

Web site: www.fda.gov

The FDA is charged with protecting the American public from unsafe foods, drugs, and cosmetics. It approves the use of antidepressants and other prescription medications and issues advisories about their use. It publishes the *FDA Consumer* magazine.

For Further Research

Books

Ava T. Albrecht and Charles Herrick, *100 Questions & Answers About Depression*. Sudbury, MA: Jones and Bartlett, 2006.

William S. Appleton, *The New Antidepressants and Antianxieties*. New York: Penguin, 2004.

Samuel H. Barondes, *Better than Prozac: Creating the Next Generation of Antidepressant Drugs*. New York: Oxford University Press, 2003.

Hyla Cass, *St. John's Wort: Nature's Blues Buster*. Garden City Park, NY: Avery, 1998.

William Dudley, ed., *Antidepressants*. Farmington Hills, MI: Greenhaven, 2005.

Kitty Dukakis with Larry Tye, *Shock: The Healing Power of Electroconvulsive Therapy*. New York: Avery, 2006.

Henry Emmons with Rachel Kranz, *The Chemistry of Joy*. New York: Fireside, 2006.

Joseph Glenmullin, *The Antidepressant Solution*. New York: Free Press, 2005.

David Healy, *Let Them Eat Prozac: The Unhealthy Relationship Between the Pharmaceutical Industry and Depression*. New York: New York University Press, 2004.

Allan Horwitz and Jerome Wakefield, *The Loss of Sadness: How Psychiatry Transformed Normal Sorrow into Depressive Disorder*. New York: Oxford, 2007.

David A. Karp, *Is It Me or My Meds? Living with Antidepressants*. Cambridge, MA: Harvard University Press, 2006.

Suzanne LeVert, *The Facts About Antidepressants*. New York: Marshall Cavendish, 2007.

Jennifer Rozines Roy, *Depression*. New York: Benchmark, 2005.

Timothy E. Wilens, *Straight Talk About Psychiatrists for Kids*. New York: Guilford, 2004.

Periodicals

J.A. Bridge et al., "Clinical Response and Risk for Reported Suicide Ideation and Suicide Attempts in Pediatric Antidepressant Treatment: A Meta-Analysis of Randomized Controlled Trials," *JAMA*, April 18, 2007.

Jordana Brown, "SAMe: Brighten Your Outlook with This Alternative to Antidepressants," *Better Nutrition*, November 2005.

Benedict Carey, "Youth, Meds, and Suicide," *Los Angeles Times*, February 2, 2004.

John Cloud, "When Sadness Is a Good Thing," *Time*, August 27, 2007.

Richard DeGrandpre, "Trouble in Prozac Nation," *Nation*, January 5, 2004.

David Dobbs, "Perspectives: Antidepressants: Good Drugs or Good Marketing?" *Scientific American Mind*, December 2004.

Tony Dokoupil, "Trouble in a 'Black Box,'—Did an Effort to Reduce Teen Suicide Backfire?" *Newsweek*, July 16, 2007.

Tiffany Kary, "Are Antidepressants Addictive?" *Psychology Today*, July/August 2003.

Joshua Kendall, "Talking Back to Prozac," *Boston Globe*, February 1, 2004.

Deborah Kotz, "The Right Rx for Sadness," *U.S. News & World Report*, August 6, 2007.

_____, "When Depression Goes Untreated," *U.S. News & World Report*, August 6, 2007.

Hara Marano, "How to Take an Antidepressant," *Psychology Today*, January /February 2003.

Sara Markowitz and Alison Cuellar, "Antidepressants and Youth: Healing or Harmful?" *Social Science & Medicine*, May 2007.

Michael C. Miller, "Exercise Is a State of Mind—Researchers Are Learning More About How Physical Activity Affects Our Moods. Is Sweat the Hot New Antidepressant?" *Newsweek*, March 26, 2007.

_____, "The Next Wave of Antidepressants," *Newsweek*, December 15, 2003.

Joanna Moncrief, "Are Depressants as Effective as Claimed? No, They Are Not Effective at All," *Canadian Journal of Psychiatry*, February 2007.

M. Olfson et al., "Antidepressant Drug Therapy and Suicide in Severely Depressed Children and Adults: A Case-Control Study," *Archives of General Psychiatry*, August 2006.

Lakshmi Ravindran and Sidney H. Kennedy, "Are Antidepressants as Effective as Claimed? Yes, but . . ." *Canadian Journal of Psychiatry*, February 2007.

Allen F. Shaughnessy, "No Difference Among New Antidepressants," *American Family Physician*, January 15, 2006.

Michael E. Thase and John C. Markowitz, "Treating Depression: Pills or Talk," *Scientific American Mind*, December 2004.

Web Sites

All About Depression (www.allaboutdepression.com). Operated by Prentiss Price, a counseling psychologist, this Web site is aimed at providing accurate and current information on clinical depression.

Briefing on Drugs for Depression (http://thomasjmoore.com). This Web site by investigative journalist Thomas J. Moore features his articles that argue that antidepressants have serious side effects and little benefit.

HedWeb: Good Drug Guide (www.biopsychiatry.com). The Web site provides an analysis of many drugs, legal and illegal, that can affect people's mood, and includes discussion of many antidepressants and herbal remedies. Includes links.

Internet Mental Health (www.mentalhealth.com). Internet Mental Health is a free encyclopedia of mental health information created by Phillip

Long, a Canadian psychiatrist. It includes a section on antidepressants and other medication and lists their recommended doses, side effects, and other information.

Justiceseekers.com (www.justiceseekers.com). A Web site maintained by a group of trial lawyers who have brought wrongful death lawsuits against antidepressant manufacturers.

Mindyourmind (http://mindyourmind.ca). This award-winning Web site and Internet community features resources developed by youth for youth on mental health issues. In includes materials from the Canadian Mental Health Association.

Paxilprogress (www.paxilprogress.org). This Web site offers advice on how to get off of SSRI antidepressants.

Source Notes

Overview

1. Quoted in Donald E. Cooley, "The Drug That Awakens Energies," *Better Homes and Gardens*, October 1957.
2. Peter D. Kramer, *Listening to Prozac*. New York: Viking, 1993.
3. Suzanne LeVert, *The Facts About Antidepressants*. New York: Marshall Cavendish, 2007.
4. Nancy Wartik, "Prozac: The Verdict Is In," *American Health*, vol. 15, November 1996.
5. Edward Shorter, "How Prozac Slew Freud," *American Heritage*, September 1998.

How Safe and Effective Are Antidepressants?

6. Tracy Thompson, "The Wizard of Prozac," *Washington Post*, November 21, 1993.
7. Quoted in Erica Goode, "Antidepressants Lift Clouds, but Lose 'Miracle Drug' Label," *New York Times*, June 30, 2002.
8. Quoted in Marianne Szegedy-Maszak, ". . . but Still Sad; Antidepressants Aren't the Magic That Millions Hoped," *Los Angeles Times*, March 27, 2006.
9. Quoted in David Stipp, "Trouble in Prozac," *Fortune*, November 28, 2005.
10. American Academy of Family Physicians, "Depression, How Medicine Can Help," 2005. http://familydoctor.org.
11. Shankar Vedantum, "FDA Told U.S. Drug System Is Broken," *Washington Post*, September 23, 2006.
12. Maurice Hinchey, "Improving the Food and Drug Administration," www.house.gov.
13. Quoted in Vedantum, "FDA Told U.S. Drug System Is Broken."

Should Antidepressants Be Prescribed to Children and Adolescents?

14. Tralee Pearce, "A Second Life for Teen Depression," *Globe and Mail*, August 21, 2007.
15. Steven E. Hyman, "Improving Our Brains?" *BioSocieties*, vol. 1, 2006.
16. Quoted in Lidia Wasowicz, "Antidepressants Often Not Effective, Healthy, Safe, or Good for Children," United Press International, June 17, 2006.
17. Peter Breggin, *Antidepressant-Induced Suicidality and Violence*, report, September 14, 2004. www.breggin.com.
18. Quoted in Joyce Howard Price, "School Shooter Took Mood Altering Drug; Medical Experts Debate Role in Violence," *Washington Times*, March 25, 2005.
19. Quoted in Wasowicz, "Antidepressants Often Not Effective."
20. Quoted in Rob Waters, "A Suicide Side Effect?" *San Francisco Chronicle*, January 4, 2004.
21. Quoted in Sara Berman, "Preventing a Tragic End," *New York Sun*, June 5, 2007.
22. Quoted in Tony Dokoupil, "Trouble in a 'Black Box,'" *Newsweek*, July 16, 2007.
23. Quoted in Dokoupil, "Trouble in a 'Black Box.'"

Are Antidepressants Overprescribed?

24. Quoted in Wartik, "Prozac: The Verdict Is In."

25. Quoted in Nathan Cobb, "Prozac Maker Says It Fights Depression, but That's Not the Whole Story," *Boston Globe*, April 15, 1994.

26. Quoted in Edward Shorter, *A History of Psychiatry*. New York: John Wiley & Sons, 1997.

27. Quoted in Ceci Connolly, "Distance Sought Between Doctors and Drug Industry," *Washington Post*, January 25, 2006.

28. Margot Magowan, "Pretty in Prozac," *San Francisco Chronicle*, March 18, 2001.

29. Quoted in CNN, "CDC: Antidepressants Most Prescribed Drugs in U.S." July 9, 2007. www.cnn.com.

30. Quoted in Deborah Kotz, "The Right Rx for Sadness," *U.S. News & World Report*, August 6, 2007.

Are Alternatives to Antidepressants Effective?

31. Francis Mark Mondimore, *Adolescent Depression*. Baltimore: Johns Hopkins University Press, 2002.

32. Klause Linde et al., "St. John's Wort for Depression—an Overview and Meta-analysis of Randomized Clinical Trials," *British Medical Journal*, vol. 313, no. 7,052, August 3, 1996.

33. Richard C. Shelton et al., "Effectiveness of St. John's Wort in Major Depression," *Journal of the American Medical Association*, vol. 285, no. 15, April 18, 2001.

34. Nancy Keates, "The Unmedicated Mind," *Wall Street Journal*, July 13, 2007.

35. B.K. Puri et al., "Eicosapentaenoic Acid in Treatment-Resistant Depression," *Archives of General Psychiatry*, vol. 59, 2002.

36. Keates, "The Unmedicated Mind."

37. Dwight L. Evans and Linda Wasmer Andrews, *If Your Adolescent Has Depression or Bipolar Disorder*. New York: Oxford University Press, 2005.

38. Quoted in Deborah Kotz, "Seven Instant Mood Boosters," *U.S. News & World Report*, August 6, 2007.

39. Kotz, "Seven Instant Mood Boosters."

List of Illustrations

Index

Index

About the Author

William Dudley received his degree in English from Beloit College, Wisconsin. He is a San Diego–based writer and editor whose books include *Teen Decisions: Drugs* and *Attention Deficit/Hyperactivity Disorder*.